Theoretical Chemistry for Experimental Chemists

Kazuyoshi Tanaka

Theoretical Chemistry for Experimental Chemists

Pragmatics and Fundamentals

Springer

Kazuyoshi Tanaka
Fukui Institute for Fundamental Chemistry
Kyoto University
Kyoto, Japan

ISBN 978-981-15-7193-0 ISBN 978-981-15-7195-4 (eBook)
https://doi.org/10.1007/978-981-15-7195-4

This Springer imprint is published by the registered company Springer Nature Singapore Pte Ltd.
The registered company address is: 152 Beach Road, #21-01/04 Gateway East, Singapore 189721, Singapore

Preface

Theoretical chemistry is a branch of chemistry which treats molecules and polymers with the aid of quantum mechanics or more briefly quantum chemistry. It has long been a playground for theoretical chemists and has been considered not to provide satisfactory calculation results with chemical precision although it can afford fundamental and crucial concepts. From 1990s, however, theoretical chemistry has gradually been acquiring reliance of experimental chemists due to considerable improvement in actual calculation method in theoretical chemistry and development of computers.

Nevertheless, usefulness of today's theoretical chemistry seems to be still unfamiliar to general experimental chemists in general. There have been published indeed plenty of good monographs and reviews on the theoretical chemistry but those are more or less for theoretical chemists themselves or trainees of this field, and not necessarily for experimental chemists. This may cause some indigestion of theoretical chemistry to the experimental chemists who are actually in need of using this "tool" for interpretation of their results, design of molecular reactions, and/or fabrication of molecular devices. This is perhaps due to somewhat a touch of closed mood and jargons included in such monographs. Also, the ordinary experimental chemists are usually busy in their own research and may not be able to spend much time to learn detailed theoretical chemistry from the beginning. But in these days theoretical chemistry is getting to be more and more effective and even indispensable means as one of the essential instrumental analyses with using personal computer, work station, and so on.

Considering these points this monograph has been written so as to provide pragmatic prescriptions, which would be of assistance to experimental chemists who like to attempt to utilize theoretical chemistry for their experimental researches. The author hopes that experimental chemists eventually get accustomed to actual utilization of theoretical chemistry or computational chemistry in a pragmatic sense through this monograph.

On this occasion the author is pleased to thank Dr. Shinichi Koizumi, Ms. Taeko Sato, and Ms. Asami Komada in the Editorial Office of Springer for their kind managements and patient encouragement to continue the author's motivation.

Kyoto, Japan
June 2020

Kazuyoshi Tanaka

Contents

Chapter 1
Introduction

Abstract Herein the whole concept of this monograph and guide to each chapter are to be described at first. Emphasis will be put on the pragmatic aspects of theoretical chemistry that would hopefully be of help to experimental chemists in digestions and analyses of the experimental results, in suggestions toward the molecular design, in extraction of unknown but valuable properties of molecules, in planning of novel chemical reactions, and so on.

Keywords Theoretical-chemistry calculations · Computational chemistry · Theoretical-calculation methods · Electronic-structure analyses · Computational softwares

In this chapter, the background and aims of this monograph are first described. This monograph is intended to give tangible and active application aspects that can be drawn from theoretical chemistry particularly to the experimental chemists who are developing new materials and/or searching for novel properties of materials they handle. The focusing points are to show that these theoretical-chemistry calculations can actually help understandings of a variety of chemical phenomena including chemical reactivities and plenty of electronic properties of materials in the bottom-up manners.

Along with the above viewpoint, this monograph starts from the demonstration of utilization of the electronic-structure analyses toward such various directions, so that the readers will be able to understand fertile and ample potentials of theoretical chemistry adaptable to their own problems, especially without suffering from parade of "complex formulae". The actual plan of this monograph is as follows:

In Chap. 2, actual examples out of the theoretical calculation including the most stable molecular structures, their electronic structures and properties, and analyses of chemical reactions are afforded at first. Simple theoretical backgrounds for specific properties are to be adequately supplemented in this chapter.

In Chap. 3, basic background for computational chemistry is to be given. The readers are encouraged to read this chapter upon necessity so as to obtain general idea and terminologies in theoretical chemistry. Explanations of several theoretical-calculation methods based on the Hartree-Fock molecular orbital theory, the post Hartree-Fock theory, and the density functional theory are involved in this chapter.

K. Tanaka, *Theoretical Chemistry for Experimental Chemists*,
https://doi.org/10.1007/978-981-15-7195-4_1

Note that somewhat simpler Xα method is not involved there, since its development form is considered to coinside with the density functional theory. In order to augment the basic descriptions of the above methods, Sect. 3.6 is to be placed as the Hints toward detailed computational works.

In Chap. 4, several topics utilizing analyses of the electronic structures of molecules are provided towards understanding the actual examples for chemical phenomena and materials of recent interest based on the utilization of calculation results.

It is not supposed here that the readers make their own softwares for performing their own computations, so that description on those is not contained in this monograph. For one who needs further information on computational details, it is encouraged to consult more details in the manuals of each computational software or more specialized and professional books. There are shown considerable amount of examples of the calculation results in this monograph, the references of which are listed as a matter of course. Please note that there are also some results without references, which have been obtained by the author for the purpose of this monograph. For these calculations, the author has employed a couple of softwares (Frisch et al. 2016; Glendening et al. 2018; Dovesi et al. 2014) of his own given in the References, which he likes to acknowledge.

Finally, the author should thank Dr. Hiroyuki Fueno (Assistant professor of Graduate School of Engineering, Kyoto University) for his occasional assistance for the author's computation. The author's acknowledgment also goes to the staffs and to quite a few of the students of the author's old laboratory in Kyoto University who made collaboration works with him.

References

R. Dovesi, R. Orlando, A. Erba, C.M. Zicovich-Wilson, B. Civalleri, S. Casassa, L. Maschio, M. Ferrabone, M. De La Pierre, P. D'Arco, Y. Noël, M. Causà, M. Rérat, B. Kirtman, Crystal 14, Int. J. Quantum Chem. **114**, 1287–1317 (2014)

M.J. Frisch, G.W. Trucks, H.B. Schlegel, G.E. Scuseria, M.A. Robb, J.R. Cheeseman, G. Scalmani, V. Barone, G.A. Petersson, H. Nakatsuji, X. Li, M. Caricato, A.V. Marenich, J. Bloino, B.G. Janesko, R. Gomperts, B. Mennucci, H.P. Hratchian, J.V. Ortiz, A.F. Izmaylov, J.L. Sonnenberg, D. Williams-Young, F. Ding, F. Lipparini, F. Egidi, J. Goings, B. Peng, A. Petrone, T. Henderson, D. Ranasinghe, V.G. Zakrzewski, J. Gao, N. Rega, G. Zheng, W. Liang, M. Hada, M. Ehara, K. Toyota, R. Fukuda, J. Hasegawa, M. Ishida, T. Nakajima, Y. Honda, O. Kitao, H. Nakai, T. Vreven, K. Throssell, J.A. Montgomery, Jr., J.E. Peralta, F. Ogliaro, M.J. Bearpark, J.J. Heyd, E.N. Brothers, K.N. Kudin, V.N. Staroverov, T.A. Keith, R. Kobayashi, J. Normand, K. Raghavachari, A.P. Rendell, J.C. Burant, S.S. Iyengar, J. Tomasi, M. Cossi, J.M. Millam, M. Klene, C. Adamo, R. Cammi, J.W. Ochterski, R.L. Martin, K. Morokuma, O. Farkas, J.B. Foresman, D.J. Fox, Gaussian 16, Revision A.03, Gaussian, Inc., Wallingford CT (2016)

E.D. Glendening, K. Badenhoop, A.E. Reed, J.E. Carpenter, J.A. Bohmann, C.M. Morales, P. Karafiloglou, C.R. Landis, F. Weinhold, NBO 7.0, Theoretical Chemistry Institute, University of Wisconsin, Madison (2018)

Chapter 2
Actual Potentials of Theoretical Chemistry: What Can Be Obtained

Abstract Theoretical calculations of molecules based on quantum chemistry can afford the electronic structures of those. The electronic structure of any molecule or supermolecule based on its wavefunction is the most fundamental and important one, since that gives us the starting point of further analyses of the electronic properties, optical properties, magnetic properties, and other fundamental properties of the molecule concerned. In this chapter, several actual examples of those are to be first described prior to the detailed explanation of complicated calculation methods exhibited in Chap. 3. This plan would be useful to directly have an idea of what kind of information on molecules can actually be obtained by theoretical calculations and how they should be further expanded in each subsection.

Keywords Structural optimization · Electronic structure · Electronic property · Mechanical property · Chemical reaction

2.1 Molecular Structure

For investigation of the electronic structure of a molecule, it is usually desirable that it is acquired from the most energetically stable molecular structure often referred to as the optimized structure. This is because "stable" electronic structure ought to normally be derived from the wavefunction defined in the optimized structure. In this sense, one would normally start from the structural-optimization process for the concerning molecule. Theoretical calculation is capable of providing the optimized molecular structures both in the ground state and in rather simple excited state as illustrated in Fig. 2.1, which is to be mentioned sequentially in this section.

The optimization process is achieved by finding the minimum point on the potential energy surface (PES) of each state, also called potential hypersurface, defined by the collection of energy points corresponding to all the possible structures of a molecule, as simply illustrated in Fig. 2.2 by changing all the variables which determine the molecular structure. This optimization process is somewhat similar to finding the bottom of a certain basin in actual geography. Essential difference from the geography, however, is that the optimization process for a molecular structure is generally performed on the multidimensional PES defined as function of all the

K. Tanaka, *Theoretical Chemistry for Experimental Chemists*,
https://doi.org/10.1007/978-981-15-7195-4_2

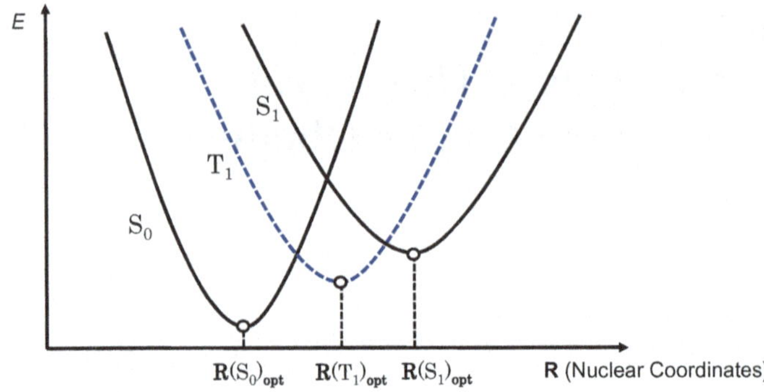

Fig. 2.1 Illustrations of the points (by white circles) indicating the optimized structures on the simplified potential energy surfaces (PES's) of the singlet ground state $\mathbf{R}(S_0)_{opt}$, the triplet excited state $\mathbf{R}(T_1)_{opt}$, and the excited singlet state $\mathbf{R}(S_1)_{opt}$

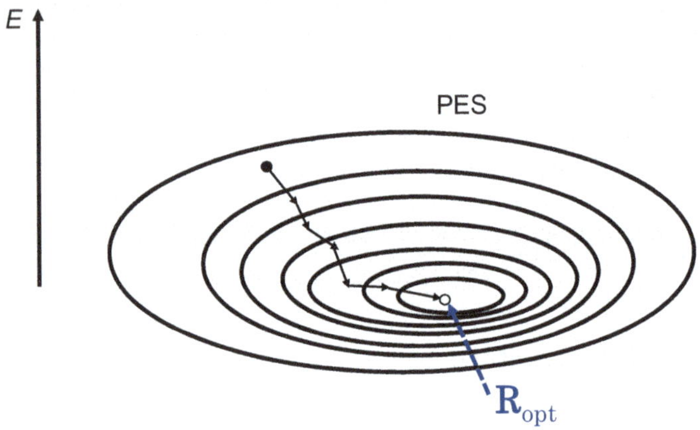

Fig. 2.2 Schematic drawing of search for the minimal point $\mathbf{R}_{opt}$ (white circle) starting from the initial point (black circle) on a PES. Note this PES can be of any state such as S_0, S_1, T_1, and so on in Fig. 2.1. Ovals represent equienergetical closed curves on the PES

variables to determine the whole molecular structure such as all the bond lengths, bond angles, and dihedral angles. In other words, the total Cartesian coordinates or their equivalences of all the nuclei are also available as the variables of a PES. The total number of these variables is equal to $3N$–5 or $3N$–6, respectively, for the linear or non-linear molecule containing N atoms by eliminating the total translation and the rotational motion of the molecules. Moreover, theoretical calculation is also capable to afford the molecular structure at the transition state (TS) during chemical reaction. This is to be mentioned in Sect. 2.7 with the actual examples.

2.1.1 Molecular Structure in the Ground State

In most cases, the structural optimization would be performed as to molecule in the ground state as described in what follows. In general, there could be at least three kinds of the initial conditions as itemized in the below for starting of the optimization process:

(1) One can utilize reliable experimental data based on, e.g., the X-ray structural analysis or its equivalence.
(2) One knows only the shape of the structure in a broad sense without knowing information on the detailed bond lengths, bond angles, nor dihedral angles.
(3) One should start from molecular formula such as $C_xH_yO_z$ without further information on the plausible structure.

In any of the above three cases for the optimization process, one ought to actually start from a certain initial geometry of the target molecule. The initial geometry is related to a certain point on the PES already shown in Fig. 2.2. Starting from the point corresponding to the initial molecular geometry, one seeks the energy-minimum point at which energy of the molecule is the most stable. A more systematic procedure of this searching for the minimum point on the PES of the ground state as well as of the excited state is to be described in Sect. 2.1.3. This optimization process and an estimation of the electronic structure of molecules obtained is now possible in most of the softwares for theoretical-chemistry calculations employing both the Hartree-Fock (HF) and the density functional theory (DFT) methods and so on. These methods are currently most popular in theoretical chemistry to treat the molecular orbitals (MO's) of molecules (see Sects. 3.1 and 3.2) with their accompanied electronic structures and electronic properties of molecules.

Even in case (1), it would be appropriate to find the optimized structure within the framework of the calculation method employed, since the electronic structure of the optimized molecular structure by each individual calculation scheme is normally desirable. It is often useful to start from the structure experimentally obtained (which is often available by the Crystallographic Information File (CIF) or its equivalence) as the initial geometry for the optimization process.

Case (2) would be usually encountered in actual optimization process. It is essentially important to start from the plausible initial geometry in order to reduce the chance of being trapped at inappropriate or undesirable molecular structures as least as possible by using any available information on the molecule and/or by relying on one's chemical intuition with respect to connectivity among atoms included. In order to build up the initial geometry, one often employs the standard bond lengths and bond angles, which are normally equipped in each calculation software or listed in the reference books and so on. There can also appear structural isomers having less energetically stable structures. Hence care should be taken as to combing out the structural isomers as exemplified in Fig. 2.3 which give several local minima on the PES.

Note that it is sometimes better to obtain the initial geometry based on a simpler calculation routine like the molecular mechanics (MM) method (see Sect. 3.4). This

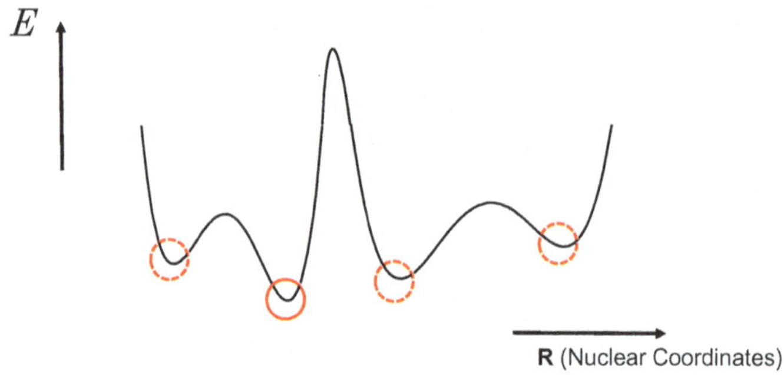

Fig. 2.3 Simple illustration of several local minimum points on the PES. The true minimum point is indicated by a solid circle

would be particularly effective when one has to deal with complicated molecular system requiring conformation analysis of oligomers or "supermolecules" consisting of large number of molecules. For comparatively small supermolecules it is possible to be treated by the optimization process of the ordinary molecules.

An example of the structure-optimized supermolecule in which several H_2 molecules are adsorbed onto an organometallic complex (Ca-catecholate) is shown in Fig. 2.4a. This structure has been obtained by theoretical calculation employing DFT/ωB97M-V/def2-TZVPD scheme that can handle molecular system or supermolecule including long-range interactions such as the van der Waals one (Tsivion et al. 2017). The notation for the calculations A/B/C signifies "calculation method/detailed scheme/basis set" throughout this book, where B (detailed scheme) is included upon necessity. This adsorption is probably caused by van der Waals

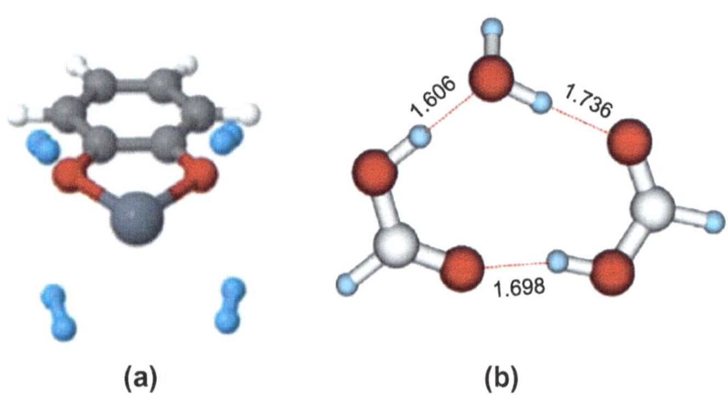

Fig. 2.4 Optimized structures of **a** calcium catecholate with four H_2 molecules adsorbed and **b** formic-acid dimer including a water molecule. Red and blue balls show oxygen and related hydrogen atoms, respectively, and black-gray one in the pentagon of **a** calcium. In **b** thin red lines represent hydrogen bonds with the distances in Å. From Tsivion et al. (2017) Copyright © 2017 by John Wiley & Sons, Inc. Reprinted by permission of John Wiley & Sons, Inc. Reprinted from Krishnakumar and Maity (2017), Copyright 2017, with permission from Elsevier

interaction including dispersion force with typical adsorption energy of the order 10–20 kJ/mol. Also note that in the "adsorption science" it is rather usual to use the energy unit of kJ/mol (= 1.036×10^{-2} eV = 3.809×10^{-4} hartree). These analyses would be useful to obtain the information on H_2 molecule storage for miscellaneous purposes such as hydrogen source for fuel cells.

Hydrogen bond is universally seen in weakly associated molecules utilizing hydrogen atoms bonded to electronegative atoms, such as N, O, or F, being particularly crucial in biomolecules. In Fig. 2.4b is shown an example consisting of a dimeric form of formic acids with a water molecule (Krishnakumar and Maity 2017) employing the DFT/ωB97X-D/aug-cc-pVDZ scheme. This system is regarded as the simplest example of microhydrated cluster of carboxylic acids. Calculated free energy of formation for the association of this cluster consisting of three hydrogen bonds at 100 K and low pressure (μTorr order) is −16.4 kcal/mol signifying a hydrogen-bond energy of 5.47 kcal/mol (=22.9 kJ/mol) in average. The hydrogen-bond energy is usually slightly larger than the van der Waals interaction energy mentioned above. For the supermolecules partly consisting of weak interaction such as van der Waals interactions or hydrogen bonds as mentioned above, one had better employ the calculation scheme such as ωB97X-D, APF-D, or CCSD(T) also applicable to these weak interactions rather than the HF or DFT/B3LYP method suitable to the ordinary covalent bonds (see Sects. 3.1 and 3.2).

Case (3) in the above might be rather rare but can happen in the design of unsynthesized molecules or when one deals with molecule the structure of which is completely unknown. In the former case, one could examine the stability against deformation of the structure of the target molecule by the structural optimization process. In the latter case, one ought to collect the experimental (spectroscopic) data such as infrared (IR) and nuclear magnetic resonance (NMR) data of the actual molecule in order to infer its structure. For the both cases, one might have to assume more than two plausible molecular structures and to perform the optimization starting from each initial geometry and compare the total energy values obtained.

In the following several examples of the optimized molecular structures are actually to be given. The optimized structure of a well-known benzene molecule C_6H_6 is shown in Fig. 2.5a. However, when one did not know the benzene structure but only knew the chemical composition C_6H_6, there can be several isomers as in Fig. 2.5b–f, each of whose energy could correspond to the local minimum point on the whole PES. In the most stable benzene with the D_{6h} symmetry, its internal degrees of freedom or, in other words, the numbers of variables are essentially two, signifying the C–C and the C–H bond lengths.

Similarly, history of clarification of the optimized structure of buckminsterfullerene, C_{60}, is somewhat of interest. The original experimental report of preparation of C_{60} showed that its mass is 720 and only gave a suggestion that this molecule would be a cluster-shape polygon with 60 carbon atoms at the vertices of a truncated icosahedron with I_h symmetry (Kroto et al. 1985). This hypothesis has actually been confirmed by the ^{13}C NMR measurement showing a single line (Taylor et al. 1990). Hence the early intuitive suggestion was indeed correct and C_{60} has the caged I_h symmetry eventually with two kinds of C–C bonds as shown in Fig. 2.6a and

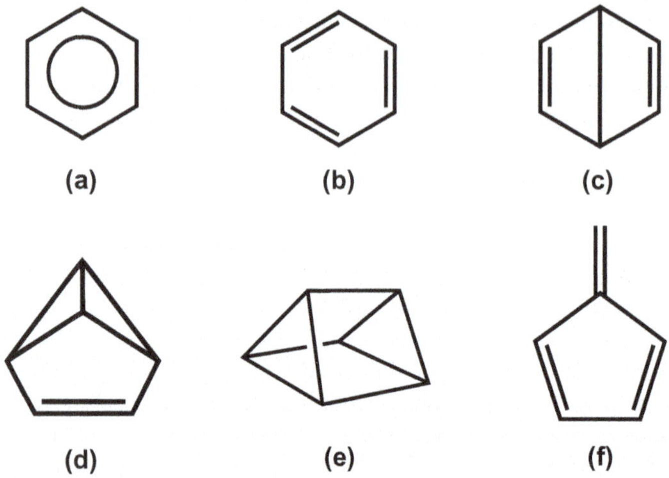

Fig. 2.5 Benzene and its isomers: **a** Benzene, **b** cyclohexatriene, **c** Dewar benzene, **d** benzvalene, **e** prismane, and **f** fulvene. Hydrogen atoms are omitted

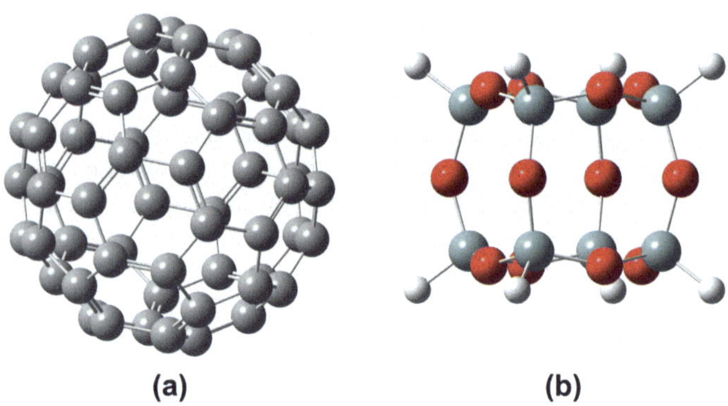

Fig. 2.6 Optimized molecular structures of **a** C_{60} and **b** POSS: $H_8Si_8O_{12}$. See Table 2.1 as to the bond lengths and angles. These were optimized by DFT/B3LYP/6-31G**

Table 2.1. Meanwhile theoretical calculations for C_{60} molecule have explosively been accumulated to explain its peculiar structure itself as well as its electronic properties (Cioslowski 1995). It is thus helpful to guess the plausible shape of molecular structure prior to or in parallel with the actual calculation as much as possible.

Moreover, the optimized structure of the modeled polyhedral oligomeric silsesquioxane (POSS) molecule $H_8Si_8O_{12}$ in Fig. 2.6b has a caged structure with O_h symmetry. This molecule is used as a nanostructured material as pendants or end-cappers to the polymers for the purpose of preventing the parent polymers from aggregation (Xiao et al. 2003; Lee et al. 2004). Moreover, the electronic properties of POSS molecule could be tuned by introducing organic functional groups connected

Table 2.1 Bond lengths and angles of C_{60} and POSS

	Bond lengths and angles[a]	X-ray diffraction[b]	Neutron diffraction[c]	Theoretical calculation[d]
C_{60}	C–C	1.455	1.46	1.453
	C=C	1.391	1.39	1.396
POSS	Si–O	1.619	1.623–1.626	1.640
	Si–H	1.450	1.459–1.463	1.460
	Si–O–Si	147.5	147.25–147.45	148.2
	O–Si–O	109.6	109.14–109.53	109.6
	O–Si–H	109.5	109.07–109.77	109.3

[a]Bond lengths in Å and angles in degrees
[b]C_{60} by David et al. (1991) and POSS by Larsson (1960)
[c]C_{60} by Li et al. (1991) and POSS by Törnroos (1994)
[d]C_{60} by the author and POSS by Mattori et al. (2000). Both the calculations by DFT/B3LYP/6-31G**

to the Si atoms (Zhen et al. 2009). Detailed structural data for POSS is also listed in Table 2.1.

The above optimization procedure of molecular structures is further applicable to the one-dimensional (1D) polymers as to their primary structures as illustrated in Fig. 2.7. This system is peculiar in that it is constructed with imposing the periodic boundary condition for the whole alignment of the unit cells providing crystal orbitals (CO's) instead of MO's. Details of the CO calculations for 1D polymer established on this condition are to be explained in Sect. 3.3. The calculation method is based on the CO concept being somewhat different from that of the usual MO. Note that in the structural optimization process the 1D polymer has one more freedom, that is, the translation length of the unit cell in addition to the numbers of the coordinate variables 3N, where N signifies the number of atoms involved in the unit cell. Note that the CO calculation is also possible for the two- and three-dimensional (2D, 3D) polymers, such as graphene and graphite crystals, respectively.

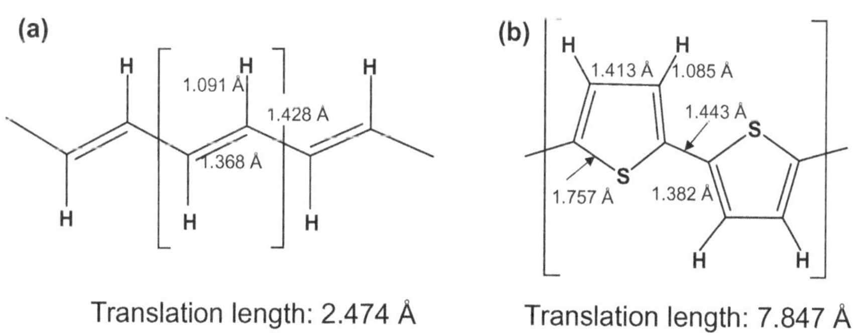

Fig. 2.7 Polymer skeletons of **a** polyacetylene and **b** polythiophene optimized by DFT-CO/B3LYP/6-31G**. The bracketed indicate the unit cells with the translation lengths

2.1.2 Molecular Structure in the Excited State

The most popular excited state dealt with by the experimental chemists would be the first singlet or the first triplet state represented as S_1 or T_1. In contrast, the singlet ground state mentioned in the above is expressed as S_0. Simple cross sections of these states have been shown in Fig. 2.1. Schematic drawings of these states are expressed by the electronic configurations shown in Fig. 2.8a–c. It is rather rare that T_1 becomes the ground state or, in other words, the existence of T_0 rarely happens except, for example, an oxygen molecule O_2 shown in Fig. 2.8d which is guaranteed by Hund's rule. O_2 molecule thus constitutes a small "magnet" in the ground state as is well known. Utilization of Hund's rule is indeed of importance for molecular design toward ferromagnetism in molecular size (see, e.g., Sects. 2.8.1 and 4.1.3).

The molecular structure in the excited state generally changes from that in the ground state. This fact is expressed by the Franck-Condon principle shown in Fig. 2.9 indicating that the vertical excitation from S_0 to S_1 state is followed by successive structural relaxation into the most stable structure in the S_1 state. In this sense, one needs to newly optimize the molecular structure in the concerning excited state. For instance, differences in geometries of formaldehyde in the S_0, S_1, and T_1 states are shown in Fig. 2.10 (Foresman et al. 1992). It is seen, for instance, the C=O bond length becomes larger in the S_1 and T_1 states implying loosening of the carbonyl bond in the excited state. On the other hand, the H–C–H angle becomes larger in the S_1 state but smaller in the T_1 state. Moreover, the C–H bond lengths do not largely change among S_0, S_1, and T_1 states. The optimization process routine for the excited state is now implemented in most of the computational softwares both for the HF and for

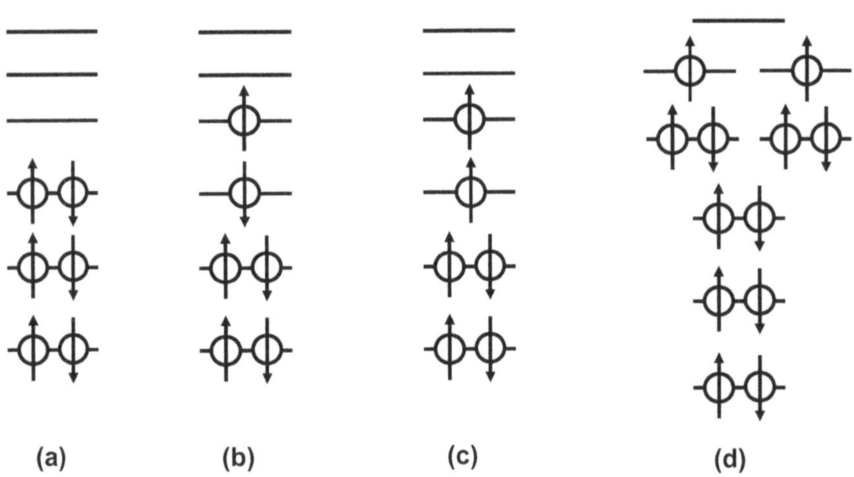

Fig. 2.8 Schematic representations of MO energy levels of **a** S_0, **b** S_1, and **c** T_1 states and **d** an O_2 molecule. The arrows with two directions indicate the α and the β spins as usual

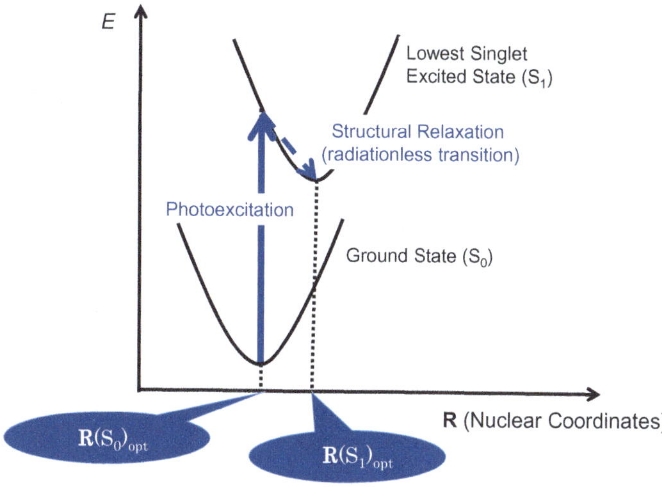

Fig. 2.9 Illustrative drawing of Franck–Condon principle, where structural relaxation takes place after the vertical photoexcitation from S_0 to S_1

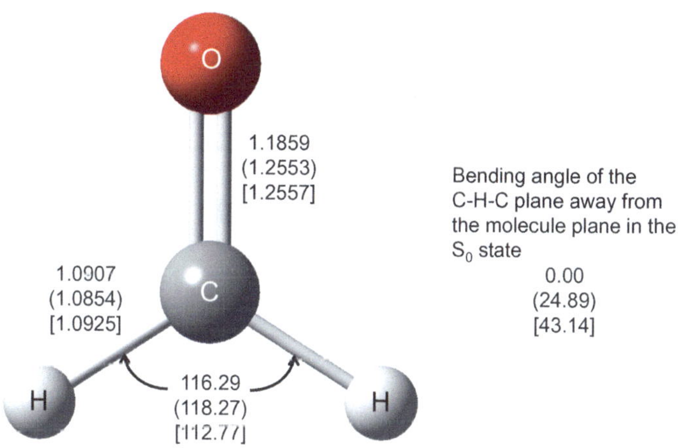

Fig. 2.10 Principal structural parameters of a formaldehyde molecule in the S_0, S_1 (in parentheses), and T_1 (in brackets) states calculated by CIS/6-31+G*. Bond lengths are in Å and bond angles in degrees. Adapted with permission from Foresman et al. (1992). Copyright 1992 American Chemical Society

the DFT methods, the latter being called time-dependent DFT (TD-DFT), which are currently most popular schemes to deal with the electronic structures and properties of molecules in the comparatively simple excited state (see Sects. 3.1 and 3.2).

The optimization process in the excited state is of particular importance in order to examine the wavelength of the photoluminescence of molecules. This is closely connected to the red shift in the photoluminescence (Stokes shift) compared with the photoexcitation spectrum due to the structural relaxation occurring in the excited state. This situation is schematically drawn in Fig. 2.11. The wavelength of the photoluminescence can be calculated as that of the photoexcitation from the S_0 to the S_1 state at the molecular structures defined by $R(S_1)_{opt}$. For instance, the data obtained along this line as 1,2-diphenylvinylene (DPV) and 1,2-diphenyldisilenylene (DPDSi) (see Fig. 2.12) are listed in Table 2.2.

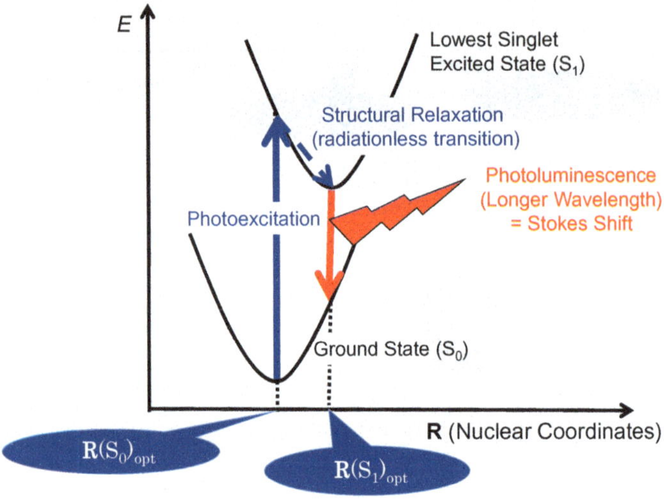

Fig. 2.11 Illustrative drawing of fluorescence from the S_1 state, which has a smaller energy than that for the photoexcitation

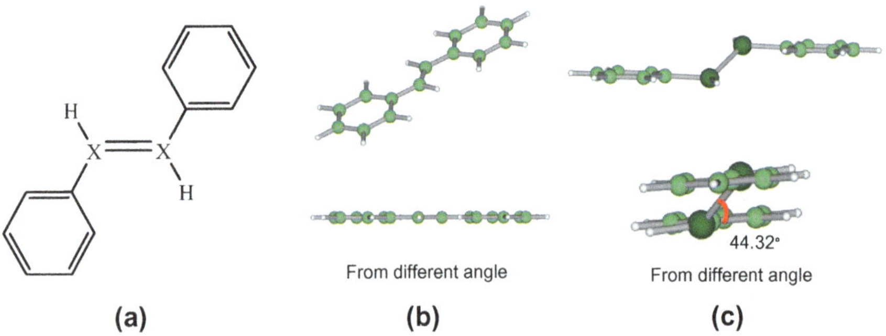

Fig. 2.12 **a** 1,2-Diphenylvinylene (DPV; X = C) and 1,2-diphenyldisilenylene (DPDSi; X = Si). In the ground state **b** structure of DPV is almost coplanar, whereas **c** that of DPDS non-planar. These were obtained by optimization at DFT/B3LYP/6-31G**

Table 2.2 Absorption and irradiation wavelengths of 1,2-diphenylvinylene (DPV) and 1,2-diphenyldisilenylene (DPDSi) between S_0 and S_1 states[a]

Compound	Absorption		Irradiation	
	Wavelength (nm)	Oscillator strength	Wavelength (nm)	Oscillator strength
DPV	309.75	0.9842	357.98	0.9792
DPDSi	400.34	0.5762	486.21	0.3978

[a]The S_0 geometry was optimized by DFT/B3LYP/6-31G** and the S_1 by CIS/6-31G**. The absorption and the irradiation wavelengths were calculated by TD-DFT/B3LYP/6-31G**

2.1.3 Stationary Points on the Potential Energy Surface (PES)

Let us mention here a simple but a bit detailed outlook of the optimization process so as to promote comprehension of relationship between the optimized structure and the normal-vibration analysis. It is currently possible to obtain the most stable molecular structure with use of many softwares by what is called the energy gradient method. Calculation based on the MM (see Sect. 3.4) is much easier and less expensive to access a plausible initial structure used in the formal optimization. The MM method, however, has a drawback in that it cannot deal with the electronic structure of molecules. Hence, in this subsection, we rather confine ourselves to the quantum chemical scheme to explicate this subject.

Finding of the optimized structure of molecule in both the ground and the excited states implies to access the "stationary point" on the PES $E(\mathbf{R})$ of each state where $\mathbf{R}$ collectively represents the $3N$ variables representing all the nuclei coordinates in the molecule consisting of N atoms. The transition state (TS) during chemical reactions is another important stationary point on the PES and will be mentioned in Sect. 2.7. Note that the degrees of freedom of the translation and rotation of the whole molecule should be removed and hence $\mathbf{R}$ essentially consists of $3N-6$ or $3N-5$ variables as has been described above. As a matter of course, $3N-6$ variables correspond to the vibrational motions of a non-linear molecule whereas $3N-5$ variables to those for linear case. But we normally include all of these degrees of freedom into the calculation for simplicity with some care. That is, the vibrational frequencies corresponding to these degrees of freedom become quite small (ideally zero) and we normally omit those. Note that combination of all the bond lengths, bond angles, dihedral angles, and so on inside the molecule can be another representative variables $\mathbf{R}$ as the "internal coordinates" for $E(\mathbf{R})$ apart from the Cartesian ones of the nuclei in the molecule.

The PES $E(\mathbf{R})$ is expressed by summation of the total electronic and the total internuclear energies of a molecule, both of which are functions of the nuclear position represented by $\mathbf{R}$. In association with the process to find out the stationary points illustrated in Fig. 2.3 on the PES, it is necessary for principle to get all the points on the PES, which clearly takes huge time and efforts. Instead, obviously from the present purpose, it is not necessary to obtain the whole PES but is enough to get

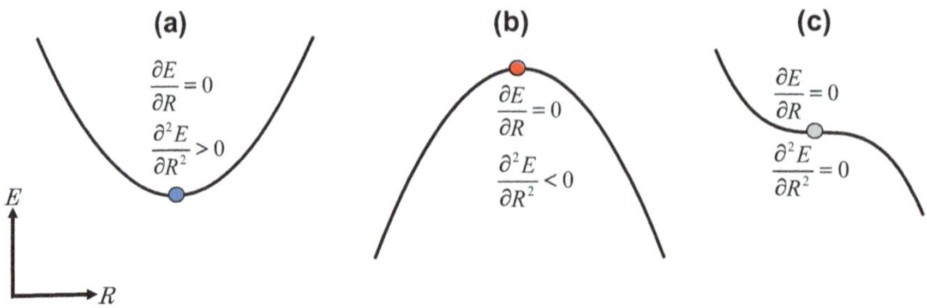

Fig. 2.13 Schematic drawing of **a** local minimum, **b** local maximum, and **c** inflection points (circles) of PES with the first and the second derivatives there in one-dimensional picture. Variable R signifies the nuclear coordinate

only the stationary points on the surface. This is because the stationary points are associated with either the local minimum or the local maximum as illustrated in Fig. 2.13a, b. Note that there is also possibility of inflection point as in Fig. 2.13c. The local minimum possibly corresponds to the most stable molecular structure or, in other words, the optimized structure, and the local maximum possibly to the TS under the condition as the saddle point on the PES as mentioned in Sect. 2.7.

In order to judge the stationary points on the PES $E(\mathbf{R})$ whether they correspond to the local minimum or local maximum, the two most important ingredients are the first and the second partial derivatives with respect to all the nuclear coordinates simply represented by R_i and R_j, that is,

$$\text{grad } E = \{\partial E/\partial R_i\}\,(i = 1, 2, \ldots, 3N) \tag{2.1}$$

and

$$\{F_{ij}\} = \{\partial^2 E/\partial R_i \partial R_j\}\,(i, j = 1, 2, \ldots, 3N) \tag{2.2}$$

where grad E signifies the gradient of the PES $E(\mathbf{R})$ with respect to a nuclear coordinate R_i, and where $\{F_{ij}\}$ constructs a force constant matrix also called Hessian. Stationary point, where grad E becomes zero, signifies either of the local minimum, local maximum, or inflection point as seen in Fig. 2.13.

It is well known that the non-zero component of –grad E implies the force along the corresponding nuclear coordinate R_i. Hence in order to reach the local minimum the nuclear displacement considering the direction and magnitude of each force shall be made with respect to all the degrees of freedom of the concerning molecule. Whether the stationary point thus obtained corresponds to the local minimum should be confirmed by examining the both of (i) convergence of all the grad E components to zero within an appropriate error range and (ii) positive values of all the eigenvalues obtained by the diagonalization of the Hessian matrix made up with the second derivatives. These conditions can be readily understood by Fig. 2.13a. The latter check is also called normal-vibration analysis. This naming comes from that the

3N–6 (or 3N–5) eigenvalues of the Hessian matrix correspond to the force constants $\{k_i\}$ for molecular normal vibrations regarded as the harmonic vibrations with the relationship

$$(2\pi\nu_i)^2 = \sqrt{\frac{k_i}{m}} \tag{2.3}$$

where $\{\nu_i\}$ stands for the frequencies of the corresponding i-th normal vibrations and m the reduced mass for the corresponding normal vibration. Thus at the local minimum all of the force constants k_i's are positive affording the "real" frequencies and the optimization process is mostly performed along with the normal-vibration analysis. Moreover, the thermochemical data such as enthalpy, entropy, and Gibbs free energy also can be obtained by making up the corresponding partition functions as well. Figure 2.14, for instance, affords the normal-vibration modes obtained at the same time of the optimization of formaldehyde molecule.

In order to effectively perform the processes described above, information on the analytical forms of $\partial E/\partial R_i$ and $\partial^2 E/\partial R_i \partial R_j$ are obviously desirable. It is currently possible to obtain the analytical forms both in the ground and in the first singlet and the triplet excited states (Foresman et al. 1992 and van Caillie and Amos 1999) and these prescriptions have been incorporated into most of the computational softwares.

1D polymers shown in Fig. 2.7 include one more freedom, translation length, say a, being added to all the nuclear coordinates **R** of the atoms constructing the unit cell as has already been mentioned. The optimization process of 1D polymers is also incorporated into the current major computational softwares.

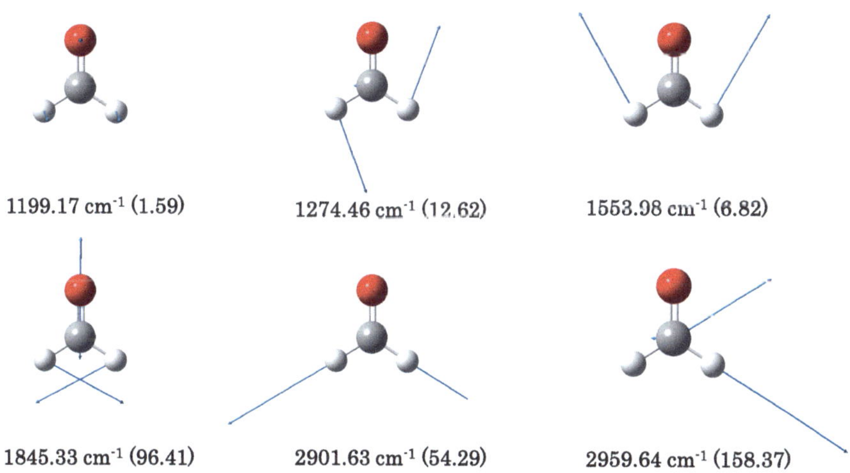

Fig. 2.14 Normal-vibration frequencies of formaldehyde in the S_0 state with the vibration modes indicated by blue arrows. IR intensities are also indicated in parentheses. These data were obtained along the optimization at DFT/B3LYP/6-31G**

2.2 Chemical Bonds

2.2.1 Concepts of Chemical Bond

Description of chemical bonds is probably most traditional and crucial subject in chemistry, which makes chemical characteristics and behavior of molecules graspable and understandable at a sight. Even in these days what is called Lewis structure based on the octet rule is considered tractable for the nature of covalent bonds and lone pairs in organic and inorganic molecules. But such ideas of chemical bonds actually might be too simple and seem to require a bit more sophisticated consideration from the quantum chemical viewpoint.

This is because the wavefunction afforded by the usual MO paradigm is not only to remain in the regions relating to each chemical bond but more or less extends to whole the molecule giving more complicated pictures than simple chemical bonds. This feature comes from the difference between the pictures given by the Lewis structure and by the MO theories. The former is directly based on the concept of the particular chemical bonds between the atoms but the latter comes from the linear combination of atomic orbitals (LCAO; AO represents the atomic orbital) formally centered at all the atoms in the molecule. Considering these situations, it is an interesting problem to reconcile these contradictory pictures from viewpoints of theoretical chemistry. In this section, the framework toward proper considerations of the chemical bonds in terms of quantum chemistry is to be explained.

2.2.2 Description of Chemical Bond

The ordinary MO's obtained by the usual calculation method, sometimes formally mentioned as the canonical MO's (CMO's), do not necessarily afford the straightforward pictures of chemical bonds in the sense of the Lewis structures. For instance, patterns of the highest occupied MO (HOMO) and the HOMO-1 of the butane CMO's shown in Fig. 2.15a are of the σ-type extending to almost all of the molecule and do not directly indicate each C–C nor C–H bond. Moreover, those of butadiene in Fig. 2.15b are of rather familiar delocalized π-type and do not afford the picture of "each" π-type bond.

In turn it is rather convenient to utilize the concept of the localized MO's (LMO's) obtained from the CMO's, since the LMO's are the set of wavefunctions intended to be localized at each particular interatomic region as much as possible to represent the chemical bonds or lone pairs existing in the molecule in terms of a certain mathematical manipulation such as unitary transformation of the set of the occupied CMO's (Edmiston and Ruedenberg 1963; Foster and Boys 1960). Due to characteristics of the unitary transformation among the occupied MO's physical meaning of the CMO's and the LMO's are the same in total. Based on similar concept, natural bond orbitals (NBO's) being a kind of LMO's are frequently used in recent years (Reed

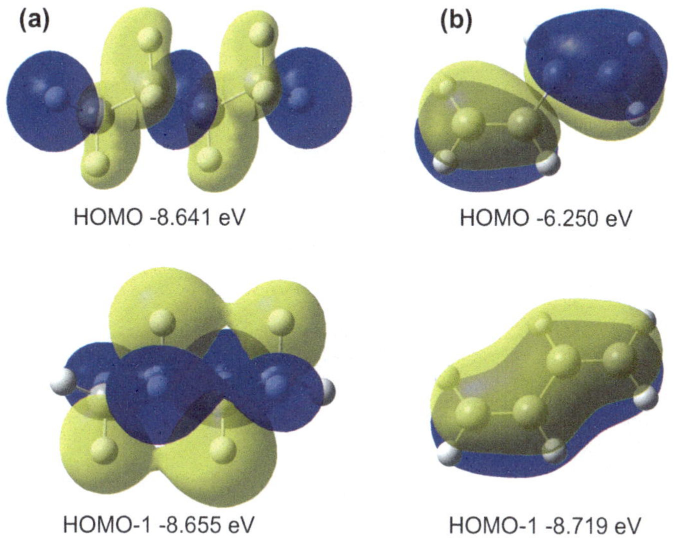

Fig. 2.15 Selected CMO patterns of **a** butane and **b** butadiene molecules with their orbital energies calculated by DFT/B3LYP/6-31G** after the structural optimization

et al. 1988). For instance, NBO's of butane and butadiene are shown in Fig. 2.16, by which specific bond pictures become somewhat clearer than those by the CMO's. That is, the two bonding NBO's of butane in Fig. 2.16a are doubly degenerated and mainly localized between the 3C and the attached two H atoms, respectively, representing two σ(C–H) bonds. There are two more degenerate NBO's localized between 2C and the attached two H atoms, which are omitted here. The two bonding NBO's

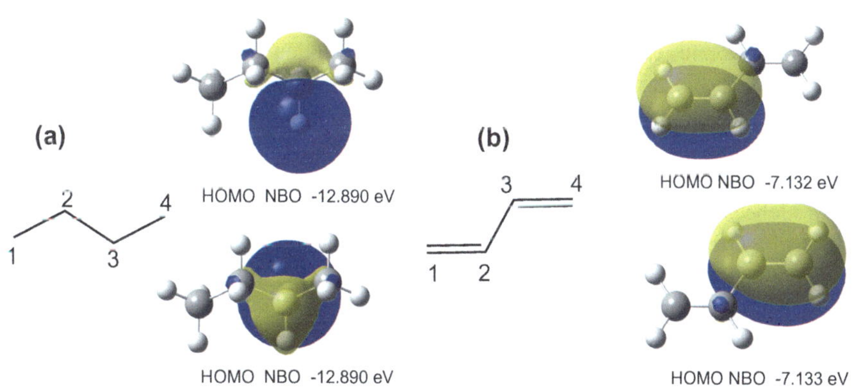

Fig. 2.16 Selected degenerated NBO patterns of **a** butane and **b** butadiene molecules with their orbital energies calculated by DFT/B3LYP/6-31G** after the structural optimization. Numberings of carbon atoms are also indicated

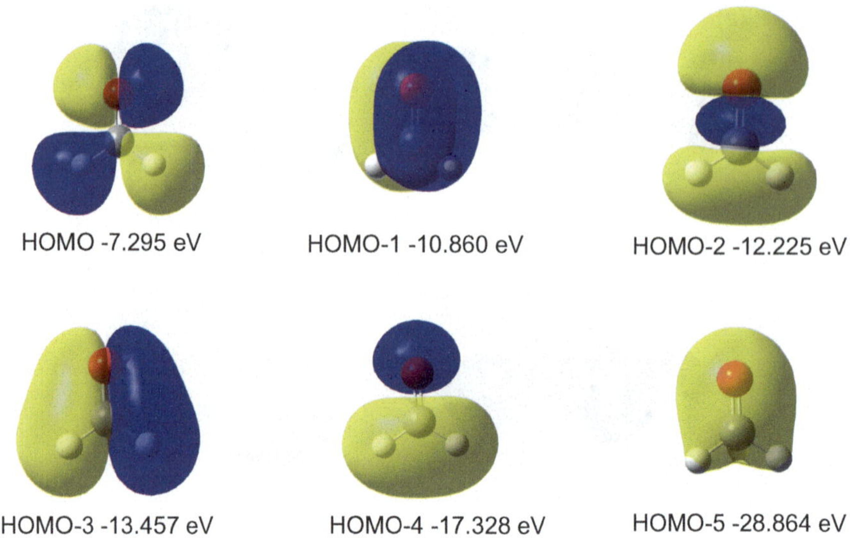

Fig. 2.17 Selected CMO patterns of formaldehyde molecule with their orbital energies calculated by DFT/B3LYP/6-31G** after the structural optimization

of butadiene in Fig. 2.16b are essentially degenerated, each representing the π(C–C) bond at 1C–2C, and 3C–4C, respectively.

NBO's are conceptually obtained from the one-particle density matrix (Löwdin 1955) by dividing that into atomic or interatomic blocks, which are appropriate to afford pictures of localized on one or two atomic centers for lone pair or chemical bond, respectively (Reed and Weinhold 1983). Lewis structure is thus clearly represented in terms of the NBO's. The CMO patterns of formaldehyde in Fig. 2.17 do not necessarily express chemical bonds except for the HOMO-1 consisting of the π(C=O) bond. On the other hand, the correspondence between the NBO's of formaldehyde and its Lewis structure is obvious as seen in Fig. 2.18. That is, for instance, two lone pairs of the O atom are clearly seen, in which one is of σ-type and the other π-type orthogonal to the O–C bond. There are also seen σ(O–C) and π(O–C) bonds and two σ(C–H) bonds. More explicit utility of NBO's is seen as four valence orbitals constructed by what is called sp^3 hybridization as shown in Fig. 2.19a. On the other hand, the CMO's of methane in Fig. 2.20a consist of triply degenerate HOMO's which are not apparently connected with the sp^3 hybridization.

A bit more sophisticated way to represent the LMO's is to utilize the NLMO (natural LMO), in which slight delocalization behavior is relegated into the antibonding NBMO's (Reed and Weinhold 1985). By this manipulation, the occupancy number of NLMO becomes just 2 whereas that of the NBO is usually a bit smaller than 2. These situations will be understood by comparison of what listed in Tables 2.3 and 2.4 describing, respectively, NBO's and NLMO's of water, for instance. The NBO's corresponding to the contents in Table 2.3 are indicated in Fig. 2.21. The computation

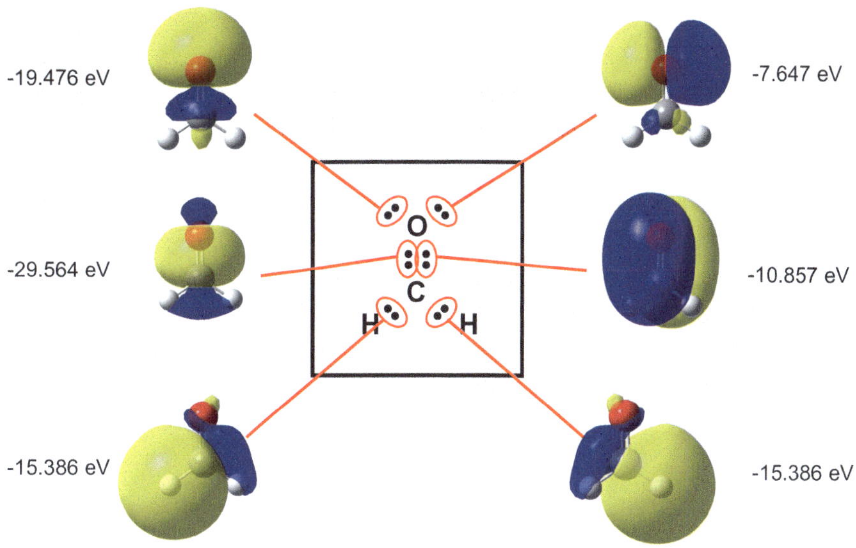

Fig. 2.18 Schematic diagram of Lewis structure and the corresponding NBO patterns of formaldehyde molecule with orbital energies calculated by DFT/B3LYP/6-31G**

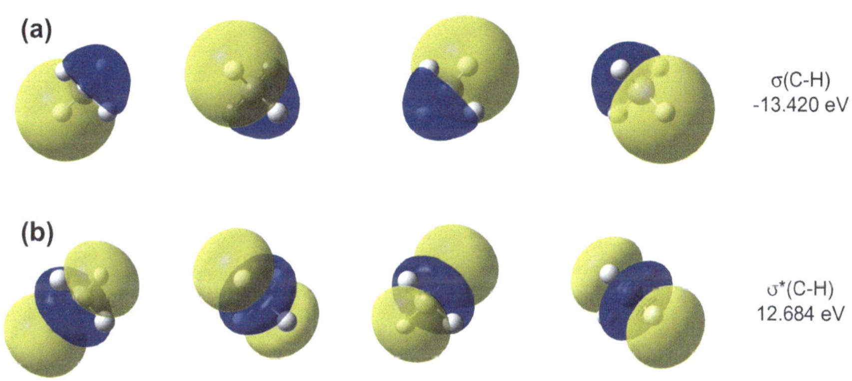

Fig. 2.19 Images of sp^3 hybridized orbitals of methane molecule by NBO with bond characteristics and orbital energies calculated by DFT/B3LYP/6-31G** after the structural optimization. **a** Bonding and **b** antibonding NBO's the both of which are quadruply degenerated

routines for the NBO and NLMO analyses are usually also implemented into most of the software packages or can be installed separately. The LMO pictures are thus convenient in understanding not only the nature of chemical bond but also inferring the chemical reaction in that they provide the direct picture related to the functional group.

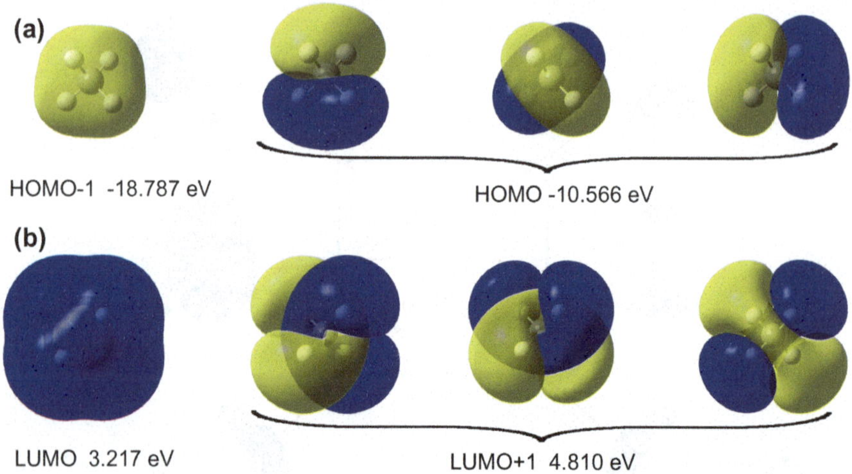

Fig. 2.20 Several CMO patterns of methane molecule with orbital symmetry and energies calculated by DFT/B3LYP/6-31G** after the structural optimization. **a** Occupied and **b** unoccupied MO's. Note that there is a node of wavefunction inside the lowest unoccupied MO (LUMO) in (**b**)

Table 2.3 NBO representation of water molecule (starting from the CMO's obtained by DFT/B3LYP/6-31G**)[a]

NBO No.	Occupancy	Orbital energy (in eV)	Characteristics
1	1.99903	−19.286	σ(O–H) $\Psi = 0.8588(sp^{3.47})_O + 0.5123s_H$
2	1.99903	−19.286	σ(O–H) $\Psi = 0.8588(sp^{3.47})_O + 0.5123s_H$
3	1.99710	−16.705	σ(lone pair at O) $\Psi = (sp^{0.80})_O$
4	1.99790	−7.868	π(lone pair at O) $\Psi = p_O$
5	0.00003	12.866	σ*(O–H) $\Psi = 0.5123(sp^{3.47})_O - 0.8588s_H$
6	0.00003	12.866	σ*(O–H) $\Psi = 0.5123(sp^{3.47})_O - 0.8588s_H$

[a]Those representing the core (2*s*) of the O atom and the Rydberg state are omitted

2.2.3 Bond Energy

The bond-energy value provides the information as to bond strength of each molecule. These values are often given by experimental technique such as mass spectrometric measurement as well as by theoretical estimations. Considerable numbers of bond-energy values are tabulated in the experimental handbook as the bond dissociation energy or dissociation enthalpies (Rumble 2018). For polyatomic molecules having

Table 2.4 NLMO representation of water molecule (starting from the CMO's obtained by DFT/B3LYP/6-31G**)

NLMO No.	Occupancy	% from the parent NBO (%)	Characteristics
1	2.00000	99.9516	σ(O–H) $\Psi = 0.8586(sp^{3.45})_O + 0.5122s_H$
2	2.00000	99.9516	σ(O–H) $\Psi = 0.8586(sp^{3.45})_O + 0.5122s_H$
3	2.00000	99.8547	σ(lone pair at O) $\Psi = 0.9993(sp^{0.80})_O + 0.0270(sp^{0.49})_H$
4	2.00000	99.8949	π(lone pair at O) $\Psi = p_O$
5	2.00000	99.9958	Core(at O) $\Psi = s_O$

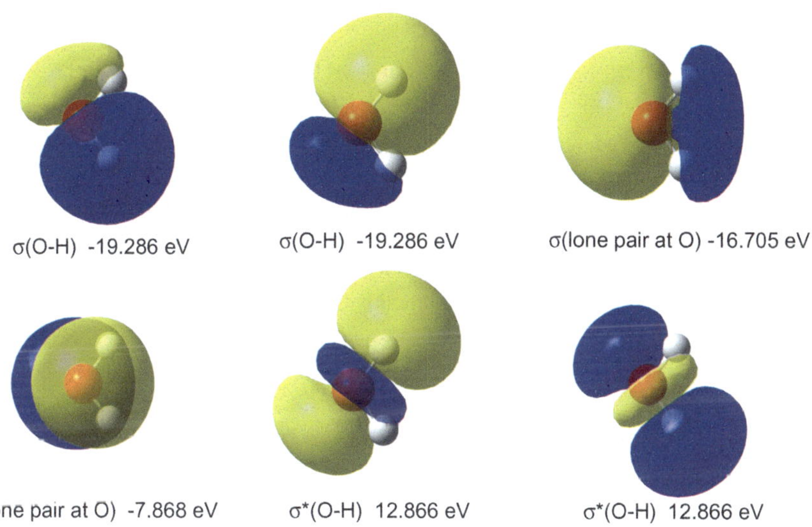

Fig. 2.21 Two bonding, two lone pairs, and two antibonding NBO patterns of water molecule with orbital energies calculated by DFT/B3LYP/6-31G** (sequentially comparable with Table 2.3)

many chemical bonds, however, the experimental measurement of bond energy becomes rather difficult. On the other hand, the theoretical enumeration of bond dissociation energies is yet possible as far as the target bond is clear.

Let us consider the energy of the bond A–B between two moieties A and B in a molecule. The most formal and straightforward definition of the bond energy E(A–B) is defined by the dissociation energy, that is,

$$E(\text{A–B}) = E(\text{AB}) - \{E(\text{A}) + E(\text{B})\} \tag{2.4}$$

where E(X) designates the energy of the molecule or its fragment X after the zero-point energy correction. Note that in this definition the bond energy is normally obtained as a negative value signifying energetical stabilization due to the bond formation. All these E(X) values can be replaced with enthalpies at certain temperature and pressure with the thermochemistry subroutine normally attached to the usual MO calculation softwares.

2.2.4 Bond Orders

Bond order is another theoretical measure for the bond strength from rather early time. This quantity is expressed by a numerical value and hence comparatively quantitative. However, one should note that there has been no decisive way of definition for the bond order. Originally, this quantity b is intuitively defined by

$$b = \frac{1}{2}\left(n - n^*\right) \tag{2.5}$$

where n and n^* designate the electron number occupying the bonding orbital and the antibonding orbital, respectively, in the sense of the LMO's. Thus the LMO's purely representing a single and a double bonds give the values of b as 1 and 2, respectively.

In a simple MO theory (Hückel MO theory) dealing with only π electrons this quantity between the sites (that is, atoms) r and s, that is, the π-AO components at r and s, is called π bond order and is defined by

$$p_{rs} = \sum_i^{\text{occ}} \nu_i c_{ir} c_{is} \tag{2.6}$$

where c_{ir} the i-th MO coefficient of the r-th π AO with ν_i being the occupation number of the i-th MO. Generally, p_{rs} becomes the total electron density between the r-th and the s-th AO's.

By inclusion of the overlap S_{rs} between the r-th and the s-th AO's the bond order p_{rs} changes into n_{rs} as

$$n_{rs} = \sum_i^{\text{occ}} \nu_i c_{ir} c_{is} S_{rs} \tag{2.7}$$

which is called AO bond population (Mulliken 1955). Collection of n_{rs} in the atoms A and B

Table 2.5 Comparison of bond orders expressed by Mulliken's atomic bond population and Wiberg's bond index (WBI) obtained by DFT/B3LYP/6-31G**)

Molecule	Bond	Mulliken's atomic bond population	Wiberg's bond index (WBI)	Remark
H_2O	O–H	0.5679	0.7767	
HCHO	O–H	0.6835	0.9123	
	C=O	1.1244	1.9295	
$H_8Si_8O_{12}$ (POSS)	Si–O	0.6846	0.5870	See Fig. 2.6b
	Si–H	0.7621	0.8817	

$$n_{\mathrm{AB}} = \sum_{r}^{\mathrm{onA}} \sum_{s}^{\mathrm{onB}} n_{rs} \tag{2.8}$$

results in the atomic bond population n_{AB}. Although these populations have been employed in the extended Hückel MO theory (Hoffmann 1963) and in the old-time HF theory, they are not frequently used in recent years.

Rather, in these days, Wiberg bond index (WBI) (Wiberg 1968; Sizova et al. 2008) for the bond between atoms A and B tends to be favored. The WBI is defined

$$\text{as} \quad W_{\mathrm{AB}} = \sum_{r}^{\mathrm{onA}} \sum_{s}^{\mathrm{onB}} p_{rs}^2 \tag{2.9}$$

where the notations are the same with those in Eq. (2.6) except that the MO coefficients employed for the calculation of p_{rs} are those obtained based on the natural AO's (see Sect. 2.3). Examples of the WBI values for a couple of simple molecules are given in Table 2.5. It is seen that there are some or considerable differences between the values of Mulliken AO bond population and WBI for each bond.

2.2.5 Weak Bonds

There are several categories of chemical bonds from the viewpoint of bond strength. Typical covalent bonds have the strengths ranging 30–100 kcal/mol, while there also exist weak bonds with the strengths of 1–5 kcal/mol such as hydrogen bond (H-bond) and that found in charge-transfer complex (CT-complex; also called electron donor-acceptor complex). These weak bonds often play crucial roles in the characteristic structures and chemical phenomena of various supermolecules, molecular complexes, and/or biological systems. The strengths ΔE of these bonds are defined as

$$\Delta E = E(\mathrm{A}\cdots\mathrm{B}) - \{E(\mathrm{A}) + E(\mathrm{B})\} \tag{2.10}$$

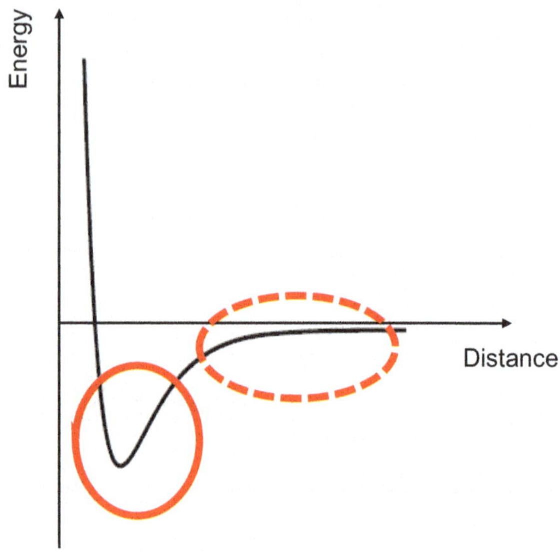

Fig. 2.22 Energy profile of the ordinary molecule with covalent bond. The solid circle represents the interatomic distance of the covalent bond suitable for usual MO calculation. The broken circle roughly sketches the region of the interatomic or intermolecular distance of the weak bonds such as in supermolecues, which requires special care (see text)

by extending Eq. (2.4) where $E(\mathrm{A} \cdots \mathrm{B})$ stands for the energy of the supermolecule $\mathrm{A} \cdots \mathrm{B}$ and hence the negative ΔE signifies stabilization upon formation of the weak bond and the positive one unstabilization.

Note that $\mathrm{A} \cdots \mathrm{B}$ distance in the supermolecule is larger than that in the normal covalent bond as seen in Fig. 2.22, which signifies that a certain care should be added to the ordinary MO calculation. Hence, by inclusion of the diffuse basis set allowing spatial extension of the wavefunctions over all the supermolecule regions and correlation effect, these weak bonds can be dealt with the MO calculation of the whole supermolecule in a successful manner toward structural optimization and estimation of the bond strength (see Sect. 3.6).

Hydrogen bond (H-bond) was originally proposed by Pauling claiming that hydrogen atom should be located between two electronegative atoms X and Y and that the hydrogen atom usually makes stronger bond with either of them (Pauling 1960). Nowadays H-bond is generally expressed by a weak interaction between covalently bonded hydrogen atom to X and an electronegative atom Y represented as

$$\mathrm{X{-}H} \cdots \mathrm{Y}$$

where H becomes slightly protonic due to charge transfer from H to Y, and where the $\mathrm{H} \cdots \mathrm{Y}$ distance is longer than those in the ordinary covalent bonds but there is obviously a certain interaction between those. Typical $\mathrm{H} \cdots \mathrm{Y}$ distance obtained by the detailed calculation is in the range 1.1–2.0 Å (Grabowski 2011). Calculation data with respect to several H-bonds are tabulated in Table 2.6 (Morokuma 1977). Magnitude of the H-bond strength is roughly correlated with polarity of X–H and electronegativity of X.

Table 2.6 Hydrogen bond with the bond strength ΔE and its decomposition obtained by HF/4-31G. Energies are in kcal/mol

Proton donor X–H	Proton acceptor Y	R (X···Y) (in Å)[a]	ΔE	ES	EX	PL	CT	MIX
FH	NH_3	2.68	−16.3	−25.6	16.0	−2.0	−4.1	−0.7
FH	OH_2	2.62	−13.4	−18.9	10.5	−1.6	−3.1	−0.4
HOH	NH_3	2.93	−9.0	−14.0	9.0	−1.1	−2.4	−0.4
HOH	OH_2	2.88	−7.8	−10.5	6.2	−0.6	−2.4	−0.5
FH	FH	2.71	−7.6	−8.2	4.5	−0.4	−3.2	−0.3
H_2NH	NH_3	3.30	−4.1	−5.7	3.6	−0.6	−1.3	−0.2
H_3CH	NH_3	4.02	−1.1	−0.6	0.5	−0.3	−0.7	−0.0

Adapted with permission from Morokuma (1977). Copyright 1977 American Chemical Society
See text as to the decomposition expressions
[a]All the (X − H··· Y) bonds are linear

Weak bonds such as H-bond are often analyzed by energy decomposition into various ingredients so as to understand the "hidden" nature of the bonds. The way of decomposition is not unique but the most popular one seems to come from Morokuma's work (Kitaura and Morokuma 1976; Morokuma 1977) by decomposing the interaction energy into electrostatic (ES), exchange repulsion (EX), polarization (PL), charge transfer (CT), and higher order coupling (MIX) energies in terms of employment of the intermediary wavefunctions. Examples of the energy decomposition as to the interaction energy of H-bond are also listed in Table 2.6. The ES term is either repulsive or attractive in general and it originates in electrostatic interaction between the two moieties involving, respectively, X–H and Y under undistorted electron distribution without formation of the H-bond. In the H-bond ES becomes mostly negative signifying an attractive electrostatic interaction. The EX term signifies contribution from electron exchange between the undistorted two moieties. Value of EX is always repulsive and is also mentioned exchange repulsion term. The PL term comes from the polarization representing distortion of electron distribution in a moiety under influence of another one and vice versa without electron exchange. This distortion is caused by electron excitation from the occupied MO to the unoccupied MO within the same moiety accompanied with charge distortion therein. The PL term causes a slight stabilization and hence its value is slightly negative. The CT term originates in electron delocalization from a moiety to another, the phenomenon of which is called charge transfer. The CT is associated with the electron transfer from the occupied MO of one moiety to the unoccupied MO of the other. The CT term always results in energetical stabilization and hence its value is negative. The MIX term summarizes the difference of the total interaction energy and the summation of the decomposed energies mentioned above, and it involves all the other coupling term of various components. The absolute value of the MIX term is not large in general. It is seen from Table 2.6 contribution from ES is generally the largest and

Table 2.7 CT-complex with the interaction energy ΔE and its decomposition obtained by HF/4-31G. Energies are in kcal/mol

D–A	CT type	R(D–A) (in Å)	ΔE	ES	EX	PL	CT	MIX
$H_3N–BF_3$	$n \rightarrow \sigma^*$	1.60	−71.5	−142.3	136.3	−42.7	−52.7	29.9
$H_3N–BH_3$	$n \rightarrow \sigma^*$	1.70	−44.7	−92.9	86.9	−17.2	−27.1	5.6
$OC–BH_3$	$\sigma \rightarrow \sigma^*$ $\pi^* \leftarrow \pi$	1.63	−28.5	−60.9	98.9	−61.8	−68.3	63.6
HF–ClF	$n \rightarrow \sigma^*$	2.74	−3.4	−3.6	1.8	−0.2	−1.4	0.1
$H_3N–F_2$	$n \rightarrow \sigma^*$	3.00	−1.1	−0.8	0.6	−0.3	−0.6	0.0
$C_6H_6–F_2$	$\pi \rightarrow \sigma^*$	3.3	−0.3	−0.2	0.3	−0.0	−0.4	0.0

Adapted with permission from Morokuma (1977). Copyright 1977 American Chemical Society

CT the second largest. The energy decomposition is useful to understand the origin of various H-bonds.

There are CT-complexes consisting of combination of various electron donors (D) and acceptors (A). Calculation data of interaction energies ΔE in several CT-complexes and the energy decomposition results are listed in Table 2.7 (Morokuma 1977). Compared with the H-bond, the interaction energies are almost the same or rather larger. For instance, in the combination of $NH_3–BF_3$, ΔE reaches −71.5 kcal/mol being comparable with the usual covalent bond. Contributions from ES and CT terms are appreciable in the combination of NH_3 and borane compounds. Theoretical calculations reveal that the CT in these complexes takes place from the lone pair of amine (n) to σ-antibonding MO of BF_3 or BH_3 (σ*). It is of interest to point out that there are two ways of the CT directions in carbonyl (CO) and borane (BH_3) complex, namely, $\sigma \rightarrow \sigma^*$ from CO to BH_3, and $\pi \rightarrow \pi^*$ from BH_3 to CO. These two-way CT directions are called "donation" and "back-donation" as is often encountered in the organometallic compounds. These donations eventually make the CT term considerably larger than those with the ordinary one-way donation.

The interatomic or intermolecular van der Waals interactions range ca. 0.01 – 0.1 kcal/mol and are much weaker than the weak bonds mentioned above providing no more perceivable bond. Origin of this interaction energy is classified into a couple of ingredients such as coming from orientation, induction, and dispersion interactions. The most essential and universal van der Waals interaction consists of the dispersion term between any atoms and between molecules involving both polar and non-polar ones.

The van der Waals interaction is said to be of importance in many chemical phenomena in (i) adsorption of molecules on surface such as catalyst, zeolite, or metal organic frameworks (MOF), (ii) π-π stacking of planar molecules such as pentacene crystal, graphite, or even aggregate of carbon nanotubes, and (iii) source of miscellaneous adhesion found in stick tape, glue, or other adhesive substances. Theoretical treatment of these interactions should be treated by the perturbation method in principle, but this method has recently been also incorporated into the

Fig. 2.23 Outlook of adsorption of a water molecule above 2D-graphene (unit cell consisting of 72 C atoms). The water molecule is adsorbed at 3.80 Å above the hollow (center of hexagon moiety) with the adsorption energy of 1.464 kcal/mol. Calculated by DFT/revPBE+vdW. Adapted with permission from Ambrosetti and Silvestrelli (2011). Copyright 2011 American Chemical Society

MO calculation. For instance, DFT calculation including van der Waals correction is sometimes performed as to the estimation of adsorption of an atom or a molecule on the surface. Figure 2.23 shows an image of adsorption of a water molecule above the 2D graphene surface along this line (Ambrosetti and Silverstrelli 2011).

2.3 Electronic Structures

In addition to the chemical-bond pictures explained in Sect. 2.2, several quantities generally mentioned as electronic structures of the molecule are frequently examined so as to figure out characteristics of molecule from many aspects. These quantities include, for instance, the orbital patterns, orbital energies (MO energies), electron density, atomic net charge, and electrostatic potential and these are to be introduced in this section. Each ingredient of the electronic structures is mostly generated from wavefunction of the molecule (namely, MO) or itself as explained in the below.

2.3.1 Orbital Patterns and Energies

The most fundamental information obtained by the quantum chemical calculation of a molecule is the wavefunction in terms of MO described under the one-electron approximation employed in the usual calculation scheme. Although there are various types of the MO calculation schemes as will be described in Chap. 3, the situation is the same so far as within the one-electron picture employed in the HF and the DFT methods. The MO is expressed by the series of coefficients for the linear combinations of the basis functions originally employed for the calculation. The molecule in the

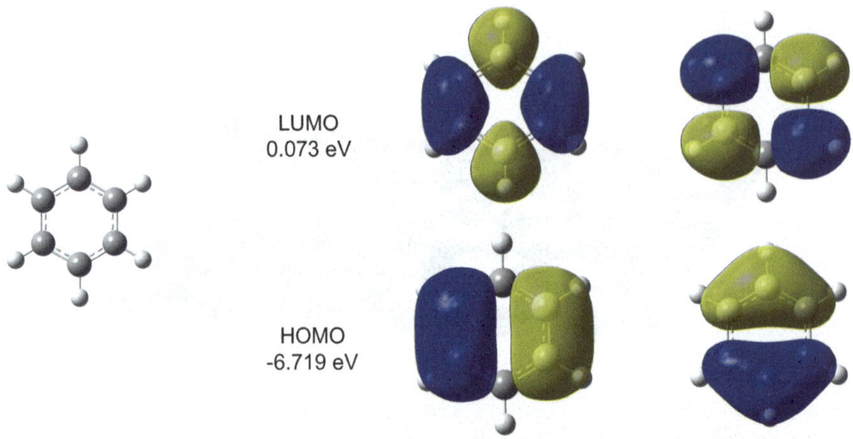

Fig. 2.24 Selected MO patterns of benzene (D_{6h} symmetry) obtained by DFT/B3LYP/6-31G** with each MO energy. Both the HOMO's and the LUMO's are doubly degenerate. The isolobe surfaces represent the MO's of the values $\pm 0.02\ e^{1/2}/au^{3/2}$ ($\pm$signs correspond to different colors) and the MO energies are indicated in eV for all the following figures unless specially noted

closed-shell ground state with $2n$ electrons, for instance, has n occupied MO's and the additional unoccupied MO's. Each MO is usually shown by the picture reflecting the magnitude of coefficients of the basis functions, which is called orbital pattern. The orbital pattern provides crucial information when one considers a variety of chemical phenomena of molecules. The orbital energy (or MO energy) accompanied by each MO signifies the energy level of the concerning MO and is shown in either au (atomic unit; also called hartree for energy) or eV unit. Both the orbital pattern and energy are of importance mostly in order to consider the reactivity and to roughly estimate the excitation energy of molecules. In particular, the patterns and the energies of the frontier MO's, that is, of the HOMO and the lowest unoccupied MO (LUMO) with their energetically neighboring MO's upon necessity, is often useful for chemists to consider chemical reactivity, photoexcitation, and other fundamental behaviors of the molecules concerned.

Examples of the MO patterns of several kinds of molecules are shown in Figs. 2.24, 2.25 and 2.26. The MO's having the same orbital energies are expressed as degenerate. For instance, both the HOMO and the LUMO of a benzene molecule are doubly degenerate as seen in Fig. 2.24. Care should be often taken to the degenerate MO's particularly when they appear in the frontier levels due to complexity of representation for the photoexcitation and so on. The MO calculation of the open-shell molecule, whose spin multiplicity is equal to or more than 2, is frequently performed with the unrestricted scheme in which the MO's for α spins and β spins are dealt with separately (DODS: different orbitals for different spins) (see Sect. 3.1). In Fig. 2.26 is shown the selected MO's of allyl radical as an example of the open-shell case. Note that the usual commercial software program assigns the number of α spins are more than that of β spins.

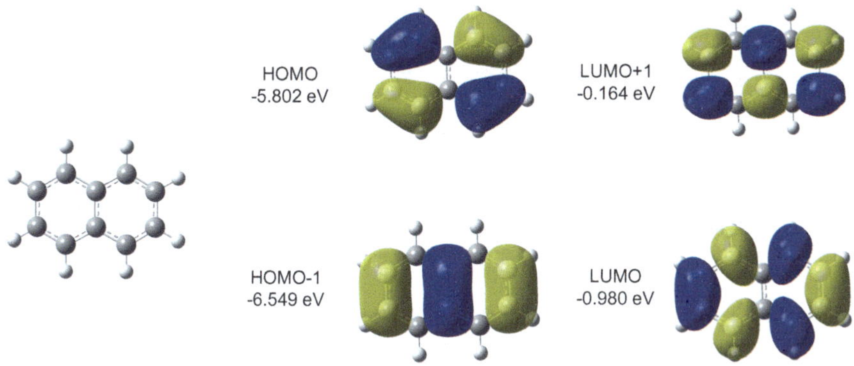

Fig. 2.25 Selected MO patterns of naphthalene (D_{2h} symmetry) obtained by DFT/B3LYP/6-31G** with each MO energy

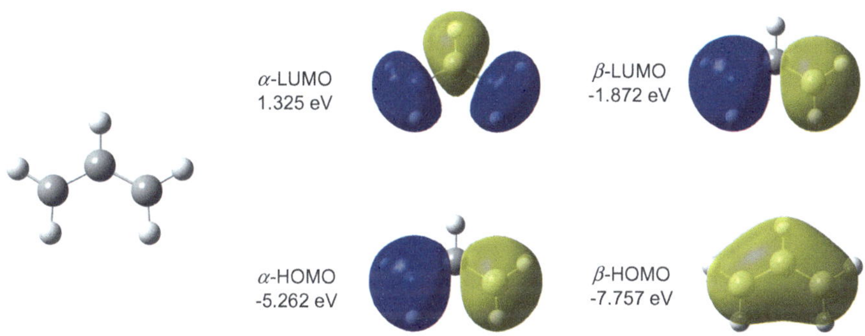

Fig. 2.26 Selected α-and and β-spin MO patterns of allyl radical (C_{2v} symmetry) obtained by unrestricted DFT/B3LYP/6-31G** with each MO energy. Note in this molecule that the patterns of α-HOMO and β-LUMO are the same in this molecule

The MO energies can be utilized to simulate the spectra obtained by the ultraviolet photoelectron spectroscopy (UPS) and/or X-ray photoelectron spectroscopy (XPS) measurements of the molecule indicating the energy levels of the valence and inner electrons in molecules. Although the MO energy is given as only a numerical value from theoretical calculations, the experimental data have normally a certain energy width whose size depends on the resolution of each spectrometer. Hence the shape of an expansion of the energy peaks from the occupied MO energy levels by the Gaussian curves, called density of valence states (DOVS), is usually compared with that of the UPS/XPS spectra. The DOVS $N(E)$ can be calculated by

$$N(E) = \sum_i \exp\left(\frac{-(E - \varepsilon_i)^2}{a}\right) \tag{2.11}$$

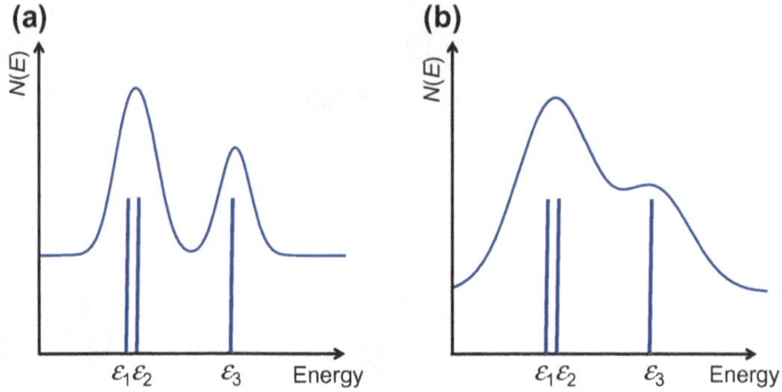

Fig. 2.27 Schematic drawings of DOVS expanded by using **a** certain two Gaussian functions and **b** those with wider widths. ε_i indicates example of MO energy

where ε_i stands for orbital energy of the i-th MO and the parameter a for the Gaussian width. The value of a can be selected by considering the energetical resolution of the spectrometer, which influences the whole shape of DOVS as schematically seen in Fig. 2.27a, b, for instance. It is thus understood that the peak separation mode becomes different according to the width of the Gaussian function. $N(E)$ can also be given as to the energy levels of the unoccupied MO's, which cannot be obtained by the UPS/XPS.

2.3.2 *Electron Density*

Summation of all the occupied MO $\psi_i(\mathbf{r})$ squared gives the total electron density $\rho(\mathbf{r})$, which is formally written by

$$\rho(\mathbf{r}) = \sum_{i}^{\text{occ}} n_i \psi_i^*(\mathbf{r})\psi_i(\mathbf{r}) \tag{2.12}$$

where n_i indicates the occupation number of $\psi_i(\mathbf{r})$ and the letter occ the summation over the occupied MO's. The asterisk indicates the complex conjugate, although the MO's are usually real functions. The total electron density is obtained by the summation of contribution from all the occupied MO's and signifies the concentration of electron cloud at the spatial position $\mathbf{r}$. Supposing the MO is represented by LCAO as usual, the MO $\psi_i(\mathbf{r})$ is written by

$$\psi_i(\mathbf{r}) = \sum_{r}^{n} c_{ir}\chi_r(\mathbf{r}) \tag{2.13}$$

with c_{ir} being the MO coefficient and n the total number of the AO's $\{\chi_r(\mathbf{r})\}$. Hence $\rho(\mathbf{r})$ in Eq. (2.12) is further decomposed into

$$\rho(\mathbf{r}) = \sum_{i}^{\mathrm{occ}} n_i \sum_{r,s}^{n} c_{ir}^{*} c_{is} \chi_r^{*}(\mathbf{r}) \chi_s(\mathbf{r}) \tag{2.14}$$

The total electron density can be represented by contour plot of the values of this quantity. For instance, the contour plots given in Figs. 2.28 and 2.29 are useful to figure out the whole electron cloud around the molecule and comparable to that obtained by the X-ray diffraction or neutron experimental data. It is seen that the electron cloud rather shrinks in OH radical and HF compared with that of HCl in Fig. 2.28. It is also clearly seen that covalencies exist in HF and HCl in a certain extent due to the presence of total electron density around the molecular axes. Moreover, in Fig. 2.29, it is seen that electron cloud is densely accumulated around oxygen in H_2O than sulfur in H_2S. This tendency is also clear in SO_2.

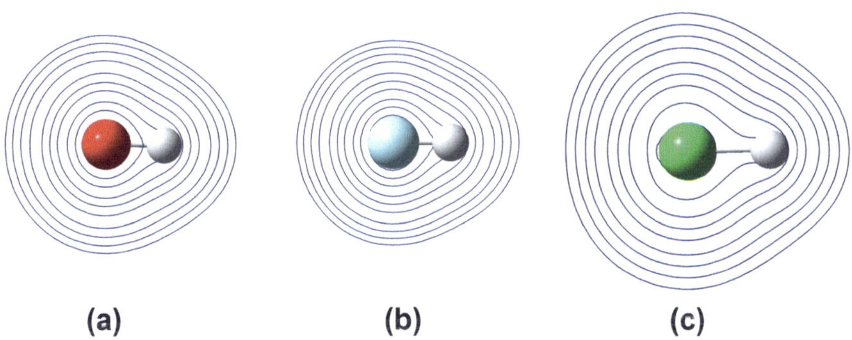

Fig. 2.28 Electron density contours of diatomic molecules: **a** OH radical, **b** HF, and **c** HCl, obtained by DFT/B3LYP/6-31G. For OH radical was used the unrestricted scheme. The outermost density contour indicates 0.0004 e/au^3

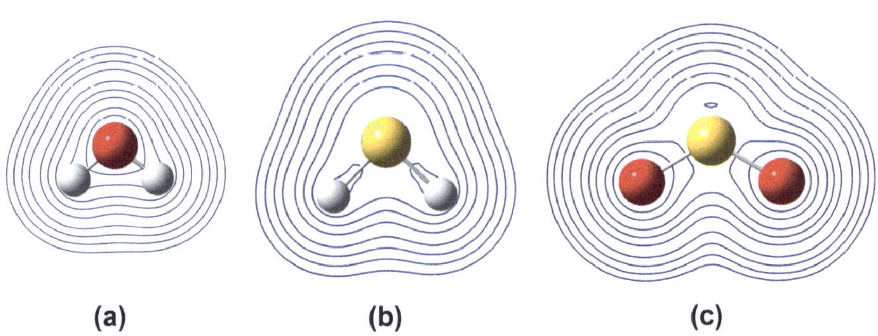

Fig. 2.29 Electron density contours of triatomic molecules: **a** H_2O, **b** H_2S, and **c** SO_2 obtained by DFT/B3LYP/6-31G**. The outermost density contour indicates 0.0004 e/au^3

2.3.3 Spin Density

In the open-shell molecule with different numbers of α and β spins, there is in general deviation of those spins depending on $\mathbf{r}$. This situation can be analyzed by considering spin density defined as a difference of α-spin density $\rho_\alpha(\mathbf{r})$ and β-spin density $\rho_\beta(\mathbf{r})$ as

$$\rho(\mathbf{r})_{\mathrm{spin}} \equiv \rho_\alpha(\mathbf{r}) - \rho_\beta(\mathbf{r}) = \sum_i^{\mathrm{occ}} n_{\alpha,i}\psi^*_{\alpha,i}(\mathbf{r})\psi_{\alpha,i}(\mathbf{r}) - \sum_j^{\mathrm{occ}} n_{\beta,j}\psi^*_{\beta,j}(\mathbf{r})\psi_{\beta,j}(\mathbf{r}) \tag{2.15}$$

In this expression, positive $\rho(\mathbf{r})_{\mathrm{spin}}$ indicates α spin is dominant at $\mathbf{r}$ and negative $\rho(\mathbf{r})_{\mathrm{spin}}$ β spin at $\mathbf{r}$. Examples of spin densities are illustrated in Fig. 2.30 showing appearance of dominant spin distribution depending on atoms. It is seen that α and β spins are shown up rather alternately, which demonstrates, what is called, spin polarization as is often seen in the open-shell molecules. Arrangement of the spin polarization of the molecules in Fig. 2.30 is indicated by arrows.

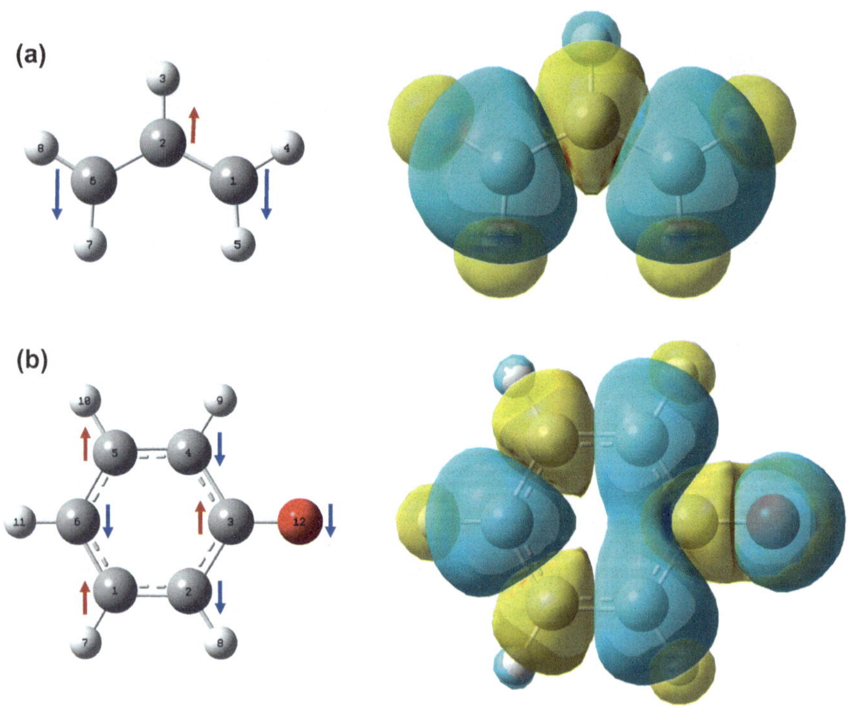

Fig. 2.30 Spin density of **a** an allyl radical and **b** phenoxy radical obtained by unrestricted DFT/B3LYP/6-31G**. Yellow and light blue colors represent the site abundant in α-spin and β-spin densities, respectively. Upward and downward arrows simply represent these spins

On the other hand, the total electron density in the open-shell molecule is obtained by

$$\rho(\mathbf{r}) \equiv \rho_\alpha(\mathbf{r}) + \rho_\beta(\mathbf{r}) = \sum_i^{\text{occ}} n_{\alpha,i}\psi_{\alpha,i}^*(\mathbf{r})\psi_{\alpha,i}(\mathbf{r}) + \sum_j^{\text{occ}} n_{\beta,j}\psi_{\beta,j}^*(\mathbf{r})\psi_{\beta,j}(\mathbf{r}) \quad (2.16)$$

which corresponds to Eq. (2.12) for the ordinary closed-shell molecule. As a matter of course, the numbers of α and β spins become the same in the singlet closed-shell molecule.

2.3.4 Atomic Net Charge

This quantity gives the measure to understand whether an atom involved in a molecule is positively or negatively charged, which provides convenient chemical intuition with experimental chemists. Atomic net charge N_{A} (A signifies an atom) is based on the information of electron population N_r from a certain AO $\chi_r(\mathbf{r})$ centered on atom A, where N_r is given by, what is called, population analysis. The atomic net charge N_{A} of the atom A can be formally expressed as

$$N_{\mathrm{A}} = Z_{\mathrm{A}} - \sum_r^{\text{on A}} N_r \quad (2.17)$$

where Z_{A} designates the nuclear charge on the atom A. It is noted that there is an essential arbitrariness for defining the electron population N_r ascribed to particular AO $\chi_r(\mathbf{r})$ since one has to cut off the electron density at a certain point although this quantity extends to the whole space. Hence this arbitrariness is also brought into atomic net charge N_{A}.

The atomic net charge, nonetheless, is a convenient and indispensable concept for experimentalists, in particular, to grasp broad idea of electronically positive or negative characteristics of each atom as well as the charge distribution over the whole molecule. This quantity is also useful for considering Lewis structure, degree of the orbital hybridizations, and so on, corresponding to the "chemical intuition" often quoted. There have been developed several ways to represent the electron population and the associated atomic net charge. In Table 2.8 are listed atomic net charges of a formaldehyde molecule, for instance, with respect to all the atoms in three ways of estimations of N_r using several kinds of basis sets as described in the following:

(1) Mulliken population analysis

Name of this analysis is after the proposer (Mulliken 1952), who is one of the founders of the MO theory in early days. This analysis is the simplest one which defines the electron population at the AO $\chi_r(\mathbf{r})$ as

Table 2.8 Mulliken and NPA (natural population analysis), AIM (atoms in molecules) atomic net charges of formaldehyde

Atom		Calc method/basis set			
		DFT/B3LYP/STO-3G	DFT/B3LYP/3-21G	DFT/B3LYP/6-31G**	DFT/B3LYP/6-311+G**
C	Mulliken	0.0211	0.0607	0.1762	0.0185
	NPA	0.1165	0.1371	0.2211	0.2896
	AIM	0.8285	0.6798	1.1034	1.0060
O	Mulliken	−0.1512	−0.3750	−0.3397	−0.2418
	NPA	−0.1535	−0.4187	−0.4945	−0.4926
	AIM	−0.8348	−0.7917	−1.1139	−1.0557
H	Mulliken	0.0650	0.1571	0.0817	0.1117
	NPA	0.0185	0.1408	0.1367	0.1015
	AIM	0.0032	0.0560	0.0053	0.0249
H	Mulliken	0.0650	0.1571	0.0817	0.1117
	NPA	0.0185	0.1408	0.1367	0.1015
	AIM	0.0032	0.0560	0.0053	0.0249

$$N_r = \sum_i^{occ} n_i \sum_s^{n} c_{ir}^* c_{is} S_{rs} \tag{2.18}$$

where S_{rs} stands for the overlap integral between AO's $\chi_r(\mathbf{r})$ and $\chi_s(\mathbf{r})$

$$S_{rs} = \int \chi_r^*(\mathbf{r})\chi_s(\mathbf{r})d\mathbf{r} \tag{2.19}$$

As is seen in Eq. (2.18) the electron population N_r comes directly from the electron density of the AO $\chi_r(\mathbf{r})$ centered on the atom A with partial contribution from that of $\chi_s(\mathbf{r})$ on other atoms through the overlap integral S_{rs}. It is noted that Mulliken population analysis is not applicable to the atoms in crystal described by the wave-function consisting of plane waves, since these are delocalized over the whole crystal and do not have particular center atom.

(2) Natural population analysis (NPA)

This analysis is a bit complicated mathematically in comparison with the Mulliken population analysis, but is claimed to be able to avoid several drawbacks of Mulliken's such as appearance of negative populations (Mulliken and Ermler 1977) and unreasonable charge distributions in ionic compounds (Collins and Streitwieser 1980).

Broad outline of the NPA is based on the following two steps (Reed et al. 1985, 1988): (i) diagonalization of the one-center (namely, an atom) localized block of the whole density matrix concerned with the AO's centered on that atom gives the eigenvectors called pre-natural atomic orbitals (pre-NAO's), and (ii) orthogonalization of thus obtained pre-NAO's is performed with respect to intra- and inter-atoms. The appropriate orthogonalization process affords the new basis set called NAO. The natural population is obtained as the diagonal element of the density matrix in the framework of the NAO basis. In other words, the NAO's are equal to the eigenfunctions of the one-center angular symmetry (like s, p, d) density matrix blocks where the natural populations are the corresponding eigenvalues. The natural population is the quantity comparable to N_r in Eq. (2.18) and can be used to afford the natural charge (NC) of the atom concerned by using Eq. (2.17). The NPA analysis and the concept of NC are frequently employed in the current computation research in molecules and the results obtained are rather accepted to the experimental chemists.

(3) Atoms in molecule (AIM) analysis

The concept of this analysis proposed by Bader (1991) has rather different aspect compared with the two kinds of analyses described in the above. In this analysis, the electron density matrix is decomposed into each atom more clearly by defining the separating surfaces on which the gradient of the electron density $\nabla\rho(\mathbf{r})$ becomes zero along the normal to the surface. This signifies the interatomic boundary is defined by that surface(s) where the electron density becomes minimum as illustrated in Fig. 2.31 (Henkelman et al. 2006). This analysis is often called atoms in molecules (AIM) method.

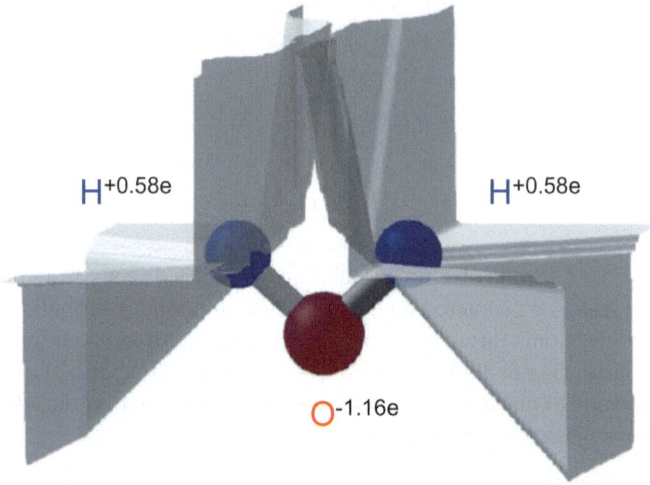

Fig. 2.31 "Bader regions" of a water molecule separated by the interatomic surfaces obtained by MP2/aug-cc-pVDZ. The integration over the Bader region gives the atomic net charges of +0.58 e of each hydrogen and −1.16 e of oxygen. Reprinted from Henkelman et al. (2006), Copyright 2006, with permission from Elsevier

Hence implication of this analysis is clear in that the atomic region, called Bader region, is uniquely defined by the surface(s) irrespective of the calculation method and selection of the basis set. On one hand, it is necessary to make efforts for numerical calculations to check the gradient of a great number of electron density grids to define the separating surfaces and then integrate these electron densities in each Bader region to get the electron population. It is emphasized that the AIM method is also applicable to estimate the atomic net charges in inorganic crystals, since in this method the basis set often can be plane waves as well. The Bader region in inorganic crystal usually becomes polyhedron.

It is seen in Table 2.8 that Mulliken population values vary depending on the basis set which appears clearly for that including polarization functions. An STO-3G basis set has a tendency of suppressing the interatomic charge flow, whereas 6-31G** not (see Sect. 3.6 with respect to these basis sets). It is further seen that AIM estimation of the atomic net charge strengthens interatomic polarization such as carbonyl whereas Mulliken and NPA estimations do not. The NPA estimation of the atomic net charge is rather moderate and this estimation is currently accepted to many experimental chemists.

2.3.5 *Electrostatic Potential*

Electrostatic potential (ESP) is defined by the potential felt by a point charge, $+1e$, which is posted as a probe on a closed surface defined by a certain

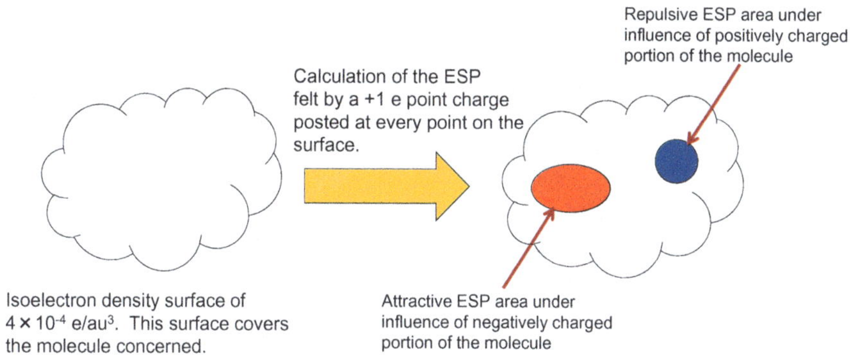

Fig. 2.32 Calculation concept to obtain the electrostatic potential (ESP)

isovalued electron density like an electron cloud around the molecule concerned as illustrated in Fig. 2.32. For instance, this potential becomes attractive on the surface area shown by reddish color under influence of negatively charged portion of the molecule, and repulsive under that of positively charged shown by bluish color. Hence function of ESP provides rather auxiliary image to the atomic net charge but is more visible to grasp the charge distribution on the whole molecule than numerical values and hence, in this sense, the ESP pictures are sometimes seen in the textbooks of organic chemistry.

A couple of examples of those are drawn in Fig. 2.33. The ESP of benzene molecule in Fig. 2.33a shows that at the center area above the benzene ring the ESP becomes attractive due to influence of the six π electrons and that the outer area where a bit positively charged hydrogens are distributed the ESP is seen to be rather repulsive. It is seen in ESP of H_2O in Fig. 2.33b that oxygen is negatively charged and hydrogens positively as a matter of course, being almost similar to sulfur and hydrogens of H_2S in Fig. 2.33c. Furthermore, for glycine in Fig. 2.33d, nitrogen and carbonyl oxygen are negatively charged, whereas hydrogen of OH group positively charged.

2.4 Electronic Properties

In this section, the electronic properties mainly originating from electronic distributions expressed by MO's and MO energies are discussed. These are inherent properties of molecule and can theoretically be examined on the basis of their electronic structures. Some are manifested as the responses to the external perturbation such as electric field. Examinations of these quantities are of particular importance when the concerning molecules are made to work as the "materials" in, e.g., actual devices or as the reagent acting in certain chemical reactions, and so on.

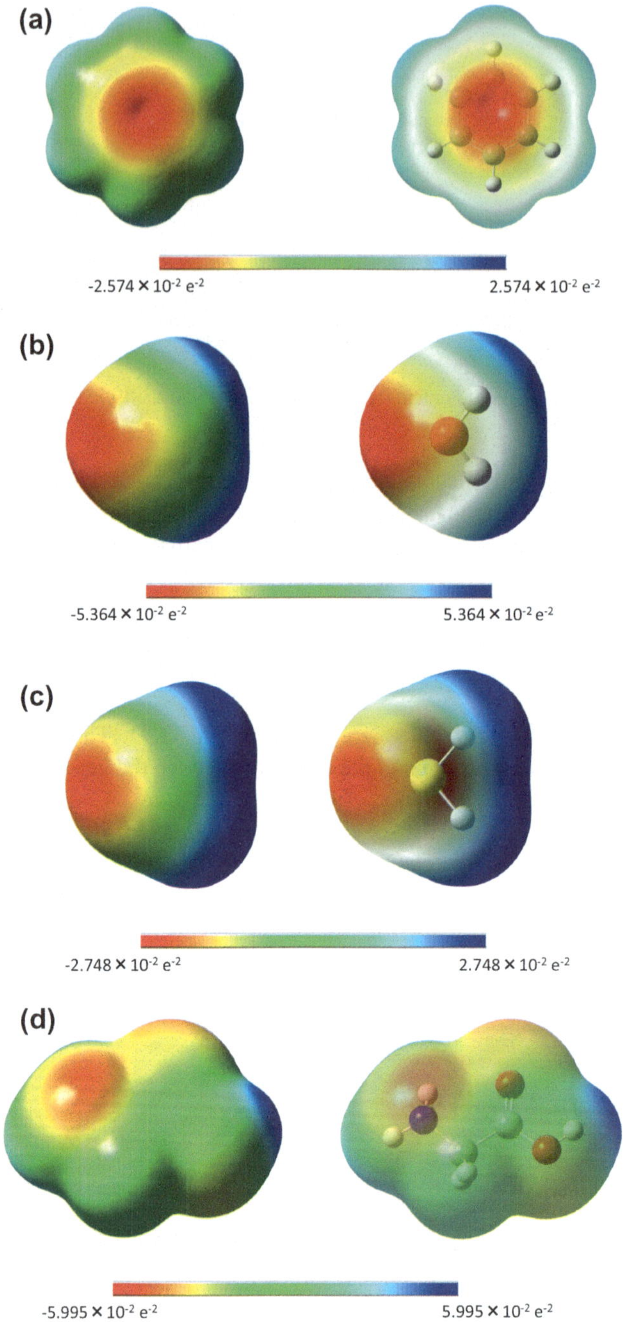

Fig. 2.33 Electrostatic potentials of **a** benzene, **b** water, **c** hydrogen sulfide, and **d** glycine with their transparent figures (each right) obtained by DFT/B3LYP/6-31G**

2.4.1 Ionization Potential and Electron Affinity

The ionization potential (I_{p}) and electron affinity (E_{A}) of a neutral molecule M are formally defined by energy difference between the neutral species M and its cationic species M^+ or anionic species M^-, respectively,

$$I_{\mathrm{p}} = E\left(\mathrm{M}^+\right) - E(\mathrm{M}) \tag{2.20}$$

and

$$E_{\mathrm{A}} = -\left\{E\left(\mathrm{M}^-\right) - E(\mathrm{M})\right\} = E(\mathrm{M}) - E\left(\mathrm{M}^-\right) \tag{2.21}$$

as illustrated in Fig. 2.34. It is noted that in general the molecular structures of M, M^+, and M^- are different as expressed in this figure. Hence $E(\mathrm{M}^+)$ and $E(\mathrm{M}^-)$ should be calculated for each optimized structures $\mathbf{R}(\mathrm{M}^+)$ and $\mathbf{R}(\mathrm{M}^-)$. Note that $E(\mathrm{M}^+)$ and $E(\mathrm{M})$ based on $\mathbf{R}(\mathrm{M}^+)$ and $\mathbf{R}(\mathrm{M})$ ought to be examined in the same calculation method and the same basis set for obtaining I_{p} and vice versa for E_{A}. Positive I_{p} requires energy injection to liberate an electron from M and negative E_{A} to add an electron to M. Hence positive E_{A} signifies that M is apt to become an anion in a spontaneous manner. The values I_{p} and E_{A}, respectively, indicate easiness of electron-donating and accepting abilities, which is closely related to the oxidation and reduction properties of species M. Experimental data of these quantities can be collected by electrochemical measurement of the redox potential or electron detachment spectroscopy. Incidentally, it is not necessary that the starting species M should be neutral. In this case, M^+ and M^- signify the charge-deviated species by + 1 and −1 from M, respectively.

As simpler ways of estimation of these two properties, one may consider the vertical process to obtain vertical I_{p} and E_{A} as also illustrated in Fig. 2.34. The vertical process means that the molecular structure $\mathbf{R}(\mathrm{M})$ does not change during the electron

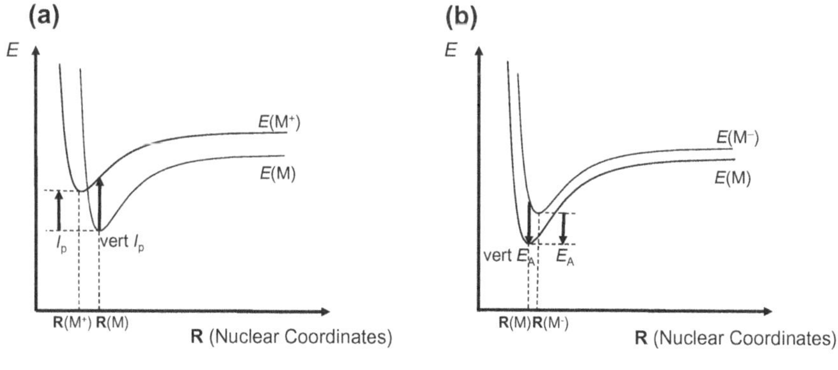

Fig. 2.34 Illustrative representation of estimating **a** I_{p} and **b** E_{A} (see text). $\mathbf{R}$(A) stands for the coordinates of the optimized structure of species A

liberation or attachment in an instantaneous interval. The structural relaxation leading to $\mathbf{R}(M^+)$ or $\mathbf{R}(M^-)$ ought to successively occur as a matter of course. Nonetheless, vertical I_p and E_A could provide reasonable measures to estimate their actual values as described in the below. Vertical I_p and E_A values are further approximated as

$$\text{vert}\quad I_p \simeq -\varepsilon_{HO} \tag{2.22}$$

$$\text{vert}\quad E_A \simeq -\varepsilon_{LU} \tag{2.23}$$

with using negatives of the HOMO and the LUMO energies. Within the framework of the ordinary MO method such as HF and DFT, equality in Eq. (2.22) holds (Koopmans 1934; Janak 1978; Perdew et al. 1982). On the other hand, Eq. (2.23) is nothing but an approximation for both the HF and the DFT methods. In general, one should include polarization and diffuse functions in the basis set (see Sect. 3.6) especially for better description of the anionic species since the excessive electron is likely to be captured in the unoccupied MO (mostly the LUMO) with spatial extension. Table 2.9 gives the DFT calculation (DFT/B3LYP/6-31+G*) data of I_p and E_A for several molecules with the available experimental values (Zhan et al. 2003). Note that the degree of approximation by Eqs. (2.22) and (2.23) is moderate in the DFT

Table 2.9 Calculated data of ionization potential (I_p), electron affinity (E_A), electronegativity (χ), and hardness (η), and negatives of the HOMO and the LUMO energies ($-\varepsilon_{HO}$ and $-\varepsilon_{LU}$) in eV obtained by DFT/B3LYP/6-31+G* with the available experimental data[a]

Molecule	I_p	E_A	χ	η[b]	$-\varepsilon_{HO}$	$-\varepsilon_{LU}$
LiH	8.283 (7.9)	0.408 (0.342)	4.346 (4.121)	3.938 (3.779)	5.322	1.318
LiCl	10.050 (10.01)	0.717 (0.59)	5.383 (5.30)	4.667 (4.71)	6.887	1.690
NaCl	9.320 (9.20)	0.873 (0.73)	5.096 (4.97)	4.224 (4.24)	6.270	2.103
CO_2	13.838 (13.773)	−0.922 (−)	6.458 (−)	7.380 (−)	10.471	0.561
H_2O	12.700 (12.621)	−2.938 (−)	4.881 (−)	7.819 (−)	8.688	−0.678
Methane	14.176 (14.40)	−1.906 (−)	6.135 (−)	8.041 (−)	10.739	−0.463
Ethylene	10.487 (10.514)	−1.772 (−)	4.357 (−)	6.130 (−)	7.546	0.218
Acetylene	11.276 (11.400)	−1.435 (−)	4.920 (−)	6.356 (−)	8.069	−0.401
Stylene	8.224 (8.43)	−0.433 (−)	3.895 (−)	4.329 (−)	6.316	1.286

[a]In parentheses are shown the experimental data from Zhang et al. (2003) and the references therein
[b]Note that in the above reference, values of 2η are actually given

framework as can be understood from the comparison of the values in Table 2.9, it is still employed at times. It has been reported that there is a certain linear relationship between the values of $|\varepsilon_i^{\mathrm{DFT}} - \varepsilon_i^{\mathrm{HF}}|$ and $\varepsilon_i^{\mathrm{HF}}$ (Stowasser and Hoffmann 1999), which has also been discussed in more detailed manner (Zhan et al. 2003). Agreement with the experimental values of the calculated values of I_p and E_A is reasonably good, while the correspondence of $-\varepsilon_{\mathrm{HOMO}}$ with I_P, and $-\varepsilon_{\mathrm{LUMO}}$ with E_A is not that good at this calculation level. This tendency signifies the approximation in Eqs. (2.22) and (2.23) ought to be utilized with certain reservation. Nonetheless, it is commonly accepted that higher HOMO level causes small I_p and lower LUMO level larger E_A.

2.4.2 Electronegativity, Hardness, and Chemical Potential

Using the I_p and E_A values, electronegativity χ (Mulliken 1934) and hardness η (Parr and Pearson 1983) of molecules are further defined as

$$\chi = \frac{I_\mathrm{p} + E_\mathrm{A}}{2} \tag{2.24}$$

$$\eta = \frac{I_\mathrm{p} - E_\mathrm{A}}{2} \tag{2.25}$$

which roughly give the auxiliary idea of electron transfer and electron-donating abilities, respectively. That is, electron transfer easily occurs from the molecule with small electronegativity χ to the other molecule with larger χ. The calculated values of χ and η are listed together with I_p and E_A in Table 2.9. The values of χ and η well reproduce the experimental ones, which might come from cancelation of errors involved in I_p and E_A values.

It is classified that an electron-donating molecule with small χ and large polarizability (see Sect. 2.4.4 as to polarizability α) tends to be easily oxidized and is called *soft base*. The molecule of large χ and small polarizability tends to be hardly oxidized being called *hard base* (Pearson 1963). In other words, *soft acid* has large χ and large polarizability, whereas *hard acid* small χ and small polarizability. Concept of hard and soft (Lewis) acids and bases (HSAB theory) has been proposed for comprehension of inorganic and organic chemical reactions (Pearson 1968) which is, however, still controversial (Mayr et al. 2011).

Along with hardness, softness S defined by

$$S = \frac{1}{2\eta} = \frac{1}{I_p - E_\mathrm{A}} \tag{2.26}$$

is also sometimes used. Employing the Eqs. (2.22) and (2.23) χ and hardness η can be approximated as

$$\chi \simeq -\frac{\varepsilon_{\mathrm{HOMO}} + \varepsilon_{\mathrm{LUMO}}}{2} \tag{2.27}$$

$$\eta \simeq \frac{\varepsilon_{\mathrm{LUMO}} - \varepsilon_{\mathrm{HOMO}}}{2} \tag{2.28}$$

signifying that η is roughly proportional to, what is called, the HOMO-LUMO energy gap.

Incidentally, chemical potential μ is defined by

$$\mu = \left(\frac{\partial E}{\partial N}\right)_V \tag{2.29}$$

with E and N being the electronic energy and the number of electrons, respectively. This concept is rather applicable to molecular aggregates or bulk including a great number of electrons, since the electronic energy E ought to be continuous for this definition. In particular, μ for solid is also called Fermi level. For semiconductive and insulating solids or polymers μ is defined as the central point of the bandgap based on the statistical mechanics for fermions. Hence μ of 1D-polymer can be written as

$$\mu = \frac{\varepsilon_{\mathrm{HOCO}} + \varepsilon_{\mathrm{LUCO}}}{2} \tag{2.30}$$

where $\varepsilon_{\mathrm{HOCO}}$ and the $\varepsilon_{\mathrm{LUCO}}$ stand for, respectively, the energies of the highest occupied CO (HOCO) and that of the lowest unoccupied CO (LUCO) (see Sect. 3.3). Extension of this idea to a molecule formally results in that the chemical potential μ of a molecule is given by

$$\mu \simeq -\chi \tag{2.31}$$

which is also of some use.

2.4.3 Dipole and Higher Moments

The dipole moment $\boldsymbol{\mu}$ is defined by

$$\boldsymbol{\mu} = \sum_{\mathrm{A}}^{\mathrm{nuclei}} (Z_{\mathrm{A}}e)\mathbf{R}_{\mathrm{A}} - e\sum_{i}^{\mathrm{elec}} \int \Psi^* \mathbf{r}_i \Psi \mathrm{d}\tau_1 \cdots \mathrm{d}\tau_N \tag{2.32}$$

where $Z_{\mathrm{A}}e$ and e are, respectively, nucleus charge of A and elementary charge as usual. $\mathbf{R}_{\mathrm{A}}$, $\mathbf{r}_i$, and Ψ are the position vector of nucleus A, that of electron i, and the total electronic wavefunction, respectively. Thus the second term of the r.h.s. of Eq. (2.32) implies the expectation value of the electron distribution of the molecule. The dipole moment is actually represented as a vector μ and stands for the overall distribution of the positive and negative charges in a molecule. This quantity relates to intermolecular force as follows: the larger |μ| gives the higher boiling point, the larger mass density, and the larger refractive index. For instance, the dipole moments of a couple of compounds are given in Fig. 2.35a–d. It is noted that there are two conventional ways of the μ vector representations depending on physics and chemistry fields as shown in Fig. 2.35e, which requires a certain attention.

The higher moments such as quadrupole, octapole, and hexadecapole ones in addition to the dipole moment can also be sequentially drawn from the multipole expansion of the electrostatic potential function as the expansion coefficients in terms of the distance between the two (positive and negative) point charges (Buckingham 1959). These higher moments are expressed as tensors and consist of the combination of dipole moments as illustrated in Fig. 2.36. In the output of the calculated

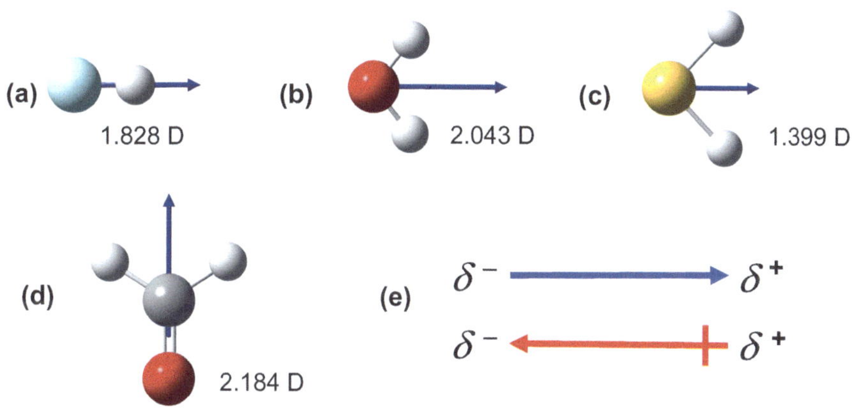

Fig. 2.35 Representation of dipole moments: **a** HF, **b** H_2O, **c** H_2S, and **d** HCHO molecules with their calculated values obtained by DFT/B3LYP/6-31G**. In **e** are shown two kinds of representations of a dipole moment in physics (upper) and chemistry (lower) fields

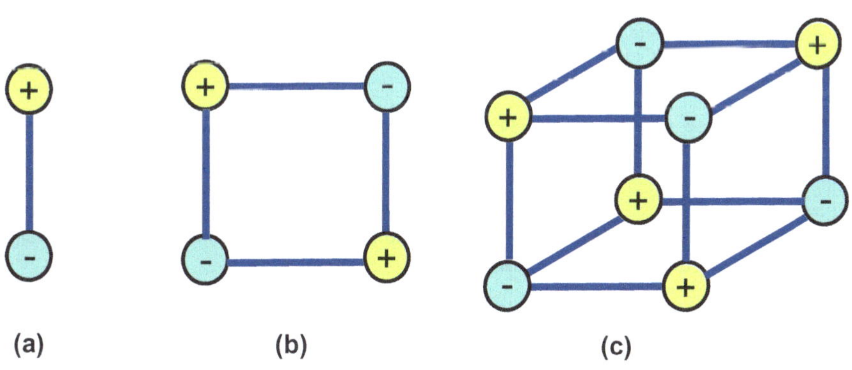

Fig. 2.36 Classical pictures of **a** dipole, **b** quadrupole, and **c** octapole moments

results of the quadrupole moment, two kinds of the representations are often seen as the primitive and the traceless versions, which comes from the different kinds of expansions. For the traceless case the spherical components are waivered, which makes the summation of *xx* and *yy* components becomes just the negative of the *zz* component. Then this traceless value corresponds to the experimentally obtained quadrupole moments data as seen in, e.g., Table 2.10. Although higher moments such as octapole and hexadecapole moments can also be obtained theoretically, the corresponding experimental values are usually unavailable. Some examples of those calculated values are given in Tables 2.10, 2.11 and 2.12.

Table 2.10 Dipole, quadrupole, and octapole moments and dipole polarizability of benzene obtained by calculation with the available experimental data

Moment		DFT/B3LYP/6-31G**	DFT/B3LYP/6-31+G*	Experimental[a]
Dipole moment **μ** (in D)	*xx*	0.0000	0.0000	–
	yy	0.0000	0.0000	–
	zz	0.0000	0.0000	–
	Tot	0.0000	0.0000	0.000
Quadrupole moment **Q** (traceless) (in D Å)	*xx*	2.3529	2.7472	2.800
	yy	2.3529	2.7472	2.800
	zz	−4.7059	−5.4944	−5.600
Octapole moment **O** (in D Å^2)	*xxx*	0.0000	0.0000	n/a
	yyy	0.0000	0.0000	
	zzz	0.0000	0.0000	
	xyy	0.0000	0.0000	
	xxy	0.0000	0.0000	
	xxz	0.0000	0.0000	
	xzz	0.0000	0.0000	
	yzz	0.0000	0.0000	
	yyz	0.0000	0.0000	
	xyz	0.0000	0.0000	
Dipole polarizability **α** (in Å^3)	*iso*	8.1477	9.7959	9.959[b]
	aniso	7.4596	5.5781	
	xx	10.6342	11.6551	
	yy	10.6343	11.6554	
	zz	3.1746	6.0772	

[a]Quadrupole moment from Flygare et al. (1971) and dipole polarizability from Gussoni et al. (1998)
[b]Corresponding to the *iso* component (α_{iso})

Table 2.11 Dipole, quadrupole, and octapole moments and dipole polarizability of formaldehyde obtained by calculation with the available experimental data

Moment		DFT/B3LYP/6-31G**	DFT/B3LYP/6-31+G*	Experimental[a]
Dipole moment **μ** (in D)	*xx*	0.0000	0.0000	–
	yy	0.0000	0.0000	–
	zz	2.1836	2.5151	–
	Tot	2.1836	2.5151	2.332
Quadrupole moment **Q** (traceless) (in D Å)	*xx*	0.2170	0.1933	−0.270
	yy	0.1161	0.2353	0.330
	zz	−0.3331	−0.4286	−0.060
Octapole moment **O** (in D Å^2)	*xxx*	0.0000	0.0000	n/a
	yyy	0.0000	0.0000	
	zzz	1.2798	0.2945	
	xyy	0.0000	0.0000	
	xxy	0.0000	0.0000	
	xxz	0.9192	0.6098	
	xzz	0.0000	0.0000	
	yzz	0.0000	0.0000	
	yyz	0.0460	−0.5382	
	xyz	0.0000	0.0000	
Dipole polarizability **α** (in Å^3)	*iso*	1.9632	2.2542	2.770[b]
	aniso	1.4966	1.5916	
	xx	1.0337	1.3182	
	yy	2.1141	2.2895	
	zz	2.7419	3.1550	

[a]Dipole moment from Hellwege and Hellwege (1982), quadrupole moment from Hellwege and Hellwege (1974), and dipole polarizability from Olney et al. (1997)
[b]Corresponding to the *iso* component (α_{iso})

2.4.4 Polarizability

The induced dipole moment $\boldsymbol{\mu}_{\text{ind}}$ under the external static electric field $\mathbf{F}$ can be expanded as

$$\boldsymbol{\mu}_{\text{ind}} = \alpha\mathbf{F} + \frac{1}{2}\beta\mathbf{F}^2 + \frac{1}{6}\gamma\mathbf{F}^3 + \cdots\cdots \quad (2.33)$$

where α is called dipole polarizability or simply polarizability, and β, γ, and so on hyperpolarizabilities (Buckingham and Orr 1967). For weak field $\boldsymbol{\mu}_{\text{ind}}$ is well represented by

Table 2.12 Dipole, quadrupole, and octapole moments and dipole polarizability of water obtained by calculation with the available experimental data

Moment		DFT/B3LYP/6-31G**	DFT/B3LYP/6-31+G*	Experimental[a]
Dipole moment $\boldsymbol{\mu}$ (in D)	*xx*	0.0000	0.0000	–
	yy	0.0000	0.0000	–
	zz	2.0428	2.2491	–
	Tot	2.0428	2.2491	1.855
Quadrupole moment **Q** (traceless) (in D Å)	*xx*	−1.3341	−1.6678	−2.500
	yy	1.5065	1.9240	2.630
	zz	−0.1724	−0.2562	−0.130
Octapole moment **O** (in D Å^2)	*xxx*	0.0000	0.0000	n/a
	yyy	0.0000	0.0000	
	zzz	−1.1613	−1.6502	
	xyy	0.0000	0.0000	
	xxy	0.0000	0.0000	
	xxz	−0.2979	−0.4919	
	xzz	0.0000	0.0000	
	yzz	0.0000	0.0000	
	yyz	−1.1939	−1.4313	
	xyz	0.0000	0.0000	
Dipole polarizability $\boldsymbol{\alpha}$ (in Å^3)	*iso*	0.7933	1.0234	1.501[b]
	aniso	0.5811	1.8742	
	xx	0.4436	0.9655	
	yy	1.1126	1.1483	
	zz	0.8236	0.9565	

[a]Dipole moment from Hellwege and Hellwege (1982), quadrupole moment from Hellwege and Hellwege (1974), and dipole polarizability from Olney et al. (1997)
[b]Corresponding to the *iso* component (α_{iso})

$$\boldsymbol{\mu}_{\text{ind}} = \alpha \mathbf{F} \tag{2.34}$$

as a matter of course. The polarizability $\boldsymbol{\alpha}$ has the form of tensor

$$\boldsymbol{\alpha} = \begin{pmatrix} \alpha_{xx} & \alpha_{xy} & \alpha_{xz} \\ \alpha_{yx} & \alpha_{yy} & \alpha_{yz} \\ \alpha_{zx} & \alpha_{zy} & \alpha_{zz} \end{pmatrix} \tag{2.35}$$

with symmetric components under arbitrary coordinate system.The isotropic $\boldsymbol{\alpha}$ signifying the mean of the diagonal components is given as

$$\alpha_{iso} = \frac{1}{3}\left(\alpha_{xx} + \alpha_{yy} + \alpha_{zz}\right) \tag{2.36}$$

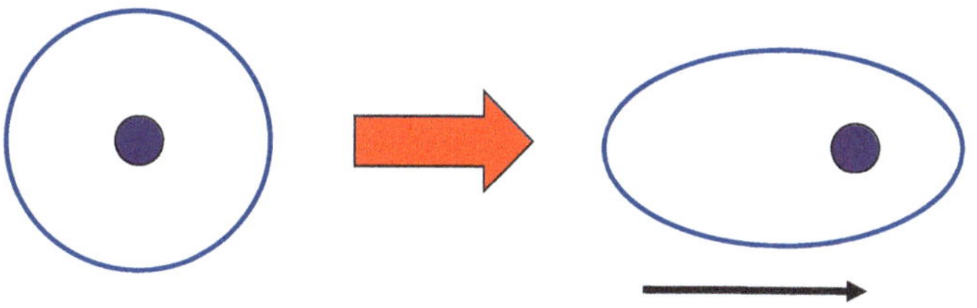

Fig. 2.37 Image of polarization of electron cloud by application of the external electric field

and the anisotropic $\boldsymbol{\alpha}$ as

$$\alpha_{aniso} = \sqrt{\frac{1}{2}\left\{\left(\alpha_{xx} - \alpha_{yy}\right)^2 + \left(\alpha_{yy} - \alpha_{zz}\right)^2 + \left(\alpha_{zz} - \alpha_{xx}\right)^2 + 6\left(\alpha_{xy}^2 + \alpha_{yz}^2 + \alpha_{zx}^2\right)\right\}} \quad (2.37)$$

These values, particularly α_{iso}, can experimentally be obtained. For the dynamic electric field changing with frequency ω the polarizabilities also become frequency dependent such as $\boldsymbol{\alpha}(\omega)$.

In Tables 2.10, 2.11 and 2.12 are also listed the data of $\boldsymbol{\alpha}$ being diagonalized so as to make $\alpha_{ij} = 0$ (for $i \neq j$). The experimental value of $\boldsymbol{\alpha}$ usually corresponds to α_{iso}. It is seen that calculated values of α_{iso} are in better agreement with the experimental ones when the basis set including diffuse functions (6-31 + G*; see Sect. 3.6 for details of the basis set). This is because the electron distribution of molecule under electric field is distorted as illustrated in Fig. 2.37, which ought to be well described by the presence of diffuse functions.

2.5 Optical Properties

Information on optical properties of molecules is of general importance since these provide the key to understand the excited state and to develop the wide application toward photochemistry, photoelectronics, and so on. Approaches in theoretical chemistry to these properties used to be insufficient due to difficulty in quantitative description of the excited state of molecules. Recently, however, considerable progress has been achieved and reliable data can be collected.

In this section, several fundamental optical properties to the experimental chemists are to be described from the viewpoints of theoretical calculation by showing actual examples. These include excitation energies and oscillator strengths, the latter of which is closely related to the absorption coefficient experimentally obtained by the optical spectra. Treatment of fluorescence and phosphorescence of organic molecules with theoretical calculations is also included since these themes could be of recent interest to the experimental chemists. Moreover, calculations related to the circular dichroism (CD) behavior of optically active molecules are to be described as well.

2.5.1 Absorption Spectrum

The most popular excited state to the experimental chemists would be the first singlet (S_1) and the triplet (T_1) excited states. Higher excited states like S_2 and T_2 are also getting to be familiar with more sophisticated purposes utilizing higher excited states. An idea of the most popular optical UV-visible (UV-vis) absorption (or electronic transition) to yield the S_1 state is schematically shown in the left side of Fig. 2.38. The single-electron excitation energy to the S_n states ($n = 1, 2, \ldots$) with the oscillator strengths accompanied is frequently estimated by the following scheme related to either the post HF or the DFT framework as in what follows:

(1) Related to the post HF: Configuration interaction singles (CI-singles or CIS) (see Sect. 3.1.2)
(2) Related to the DFT: TD-DFT (see Sect. 3.2).

In the CIS or the TD-DFT methods, the wavefunction corresponding to the S_n state is generally given by linear combination of several electronic configurations combined with multiple electron transitions employing the original MO's as, for instance,

$$\Psi_{S_n} = c_1 \Psi_{\mathrm{HO \to LU}} + c_2 \Psi_{\mathrm{HO\text{-}1 \to LU}} + c_2 \Psi_{\mathrm{HO \to LU+1}} + \cdots \tag{2.38}$$

where $\Psi_{\mathrm{HO \to LU}}$ stands for the configuration with the electron transition from the HOMO to the LUMO and so on. Equation (2.38) signifies that the wavefunction of the excited state described by the above CIS or TD-DFT automatically includes the electron correlation to a certain extent. A simple example of this representation for S_1 state is illustrated in the right side of Fig. 2.38. The most fundamental estimation

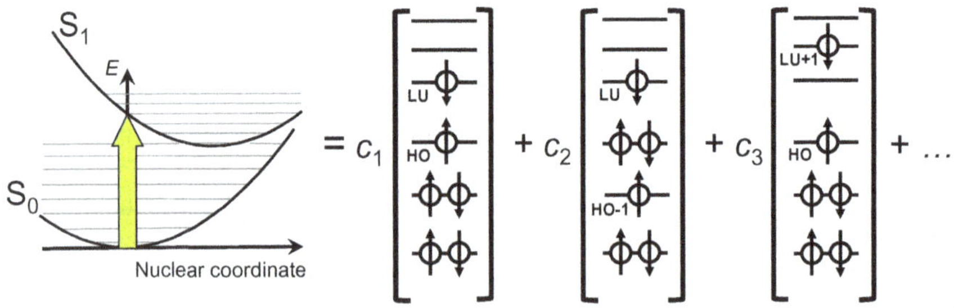

Fig. 2.38 An example of description of the single-electron excitation from S_0 to S_1 state expressed by linear combination of multiple electronic configurations

of the excitation energies by the ordinary HF scheme is also to be explained in Sect. 3.1.1.

The theoretical absorption spectrum corresponding to that obtained by the experimental measurement can be plotted by summation of the expanded Gaussian curves like density of the valence states (DOVS) in Sect. 2.3 at each excitation energy ΔE_{S_n} with the height of the corresponding oscillator strength f_{S_n}. In order to plot the calculated absorption spectrum in a realistic manner, it is generally required to consider the excitation energies with the corresponding oscillation strengths up to more than the 20th excitation. The oscillator strength value f_{S_n} is given by

$$f_{S_n} = \frac{2}{3}\Delta E_{S_n}\mathbf{m}_{S_n}^2 \tag{2.39}$$

where ΔE_{S_n} designates the excitation energy from the ground state S_0 to the excited state S_n, and $\mathbf{m}_{S_n}$ the transition moment between these two states defined by

$$\mathbf{m}_{S_n} \equiv \int \Psi_0^* \left(-e\sum_i \mathbf{r}_i\right) \Psi_{S_n} d\tau \tag{2.40}$$

with Ψ_0 and Ψ_{S_n} being the wavefunctions of the above two states, $\mathbf{r}_i$ the coordinate of the i-th electron, $d\tau$ the collective volume element of both the spatial, and the spins of the i-th electron. It is well understood that the zero value of the transition moment $\mathbf{m}_{S_n}$ signifies that the transition of $S_0 \rightarrow S_n$ is forbidden. Note that the oscillator strength f_{S_n} obtained in Eq. (2.39) is roughly proportional to the area S of molar absorption coefficient ε experimentally obtained, that is,

$$S - 4.32 \times 10^{-9} \times \int \varepsilon(\sigma) d\sigma \tag{2.41}$$

in which ε is integrated in the appropriate range of wavelength σ.

Since the optical excitation to the triplet state (T_n) is forbidden due to the integration by $d\tau$ for different spin states unless an additional interaction such as spin-orbit coupling works in the heavy-atom system, the direct transition from S_0 to T_n is normally not considered except for that by thermal process. However, the excitation energy from S_0 to T_n state could also be calculated as will be described in Sect. 3.1.1 as well. It is noted that the energy giving the apparent peaks of the absorption spectrum may be somewhat shifted from the calculated excitation energies with the large oscillator strength due to influence of the overall summation collecting the vicinal excitation energies.

In Fig. 2.39a is shown an actual example of the theoretical absorption spectrum obtained by the TD-DFT calculation for a benzotrithiophene derivative (tris(phenylisoxazolyl)benzotrithiophene) with methoxy groups attached to the outer three phenyl rings at the energetically optimized structure (Fig. 2.39b). Since it is

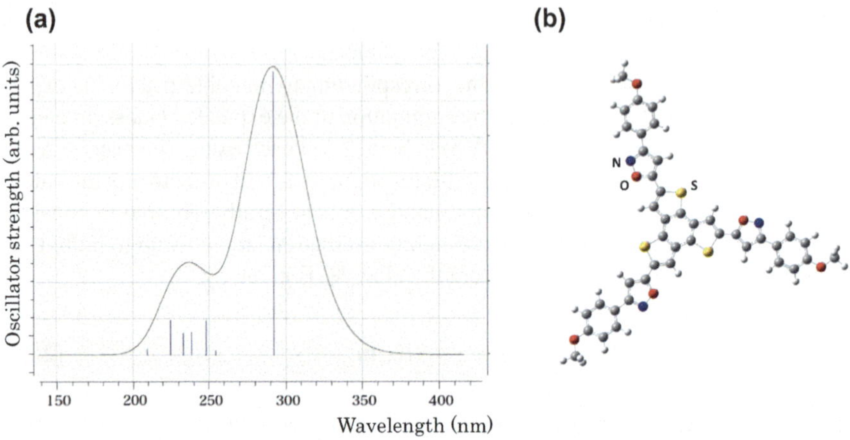

Fig. 2.39 **a** Calculated absorption spectrum at **b** the optimized structure of the benzotrithiophene derivative in vacuum (C_3 symmetry). The calculation was performed by DFT/ωB97X-D/6-31G** and the absorption data were obtained by the TD-DFT scheme (see Table 2.13)

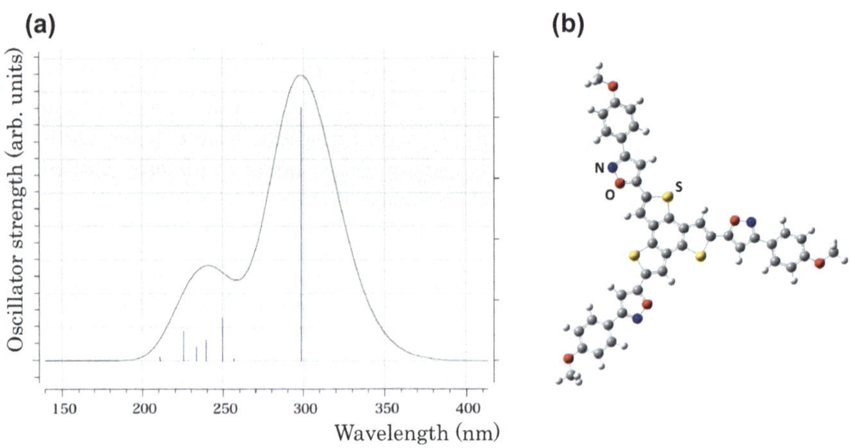

Fig. 2.40 **a** Calculated absorption spectrum at **b** the optimized structure of the benzotrithiophene derivative in methylcyclohexane (MCH) solution (C_1 symmetry). The calculation was performed by DFT/ωB97X-D/6-31G** and the absorption data were obtained by the TD-DFT scheme (see Table 2.13)

usual that UV-visi absorption measurement for molecules is performed in the solution, the calculated data for the same molecule in methylcyclohexane (MCH) solution is shown in Fig. 2.40 with the optimized molecular structure therein. The solvent effect by the conventional solvents can be incorporated into the calculation in terms of the polarizable continuum model (PCM). It is noted that the optimized structure is slightly different compared with that in vacuum (from C_3 to C_1 symmetry) and

Table 2.13 Calculated data of the main optical absorption of the benzotrithiophene derivative with the end MeO groups[a]

Condition	Excitated state	Absorption		
		Wavelength (nm)	Oscillator strength	Corresponding transitions with the coefficients
In vacuum	3	291.95	1.9517	HOMO-1 → LUMO + 1 0.3296 HOMO → LUMO 0.3296
	4	291.95	1.9517	HOMO-1 → LUMO 0.3296 HOMO → LUMO + 1 −0.3296
In MCH solution[b]	3	298.57	2.0925	HOMO-1 → LUMO 0.3024 HOMO → LUMO + 1 −0.3012
	4	298.56	2.0931	HOMO-1 → LUMO + 1 0.3029 HOMO → LUMO 0.3005

[a]The S_0 geometry was optimized by DFT/ωB97X-D/6-31G**. The absorption data were obtained by the TD-DFT scheme
[b]MCH methylcyclohexane

that the oscillator strength plot is slightly red shifted. A couple of the calculated data with large oscillator strengths are listed in Table 2.13. The experimental absorption spectrum as to similar molecule with the different end alkoxy groups ($-OC_{10}H_{21}$) in MCH solution is shown in Fig. 2.41, which shows the maximum peak at longer wavelength (Ikeda et al. 2017). It is seen that the calculation data in MCH solution tends to approach the experimental spectrum. The concerning MO patterns and their energies are shown in Figs. 2.42 and 2.43. Note that the degeneracies of the HOMO and the LUMO are released in the MCH solution due to the symmetry lowering to C_1.

2.5.2 Emission Spectrum

Next is to be examined the calculated fluorescent emission spectrum. The experimentally obtained absorption and emission spectra for a pirenyldisilene derivative are shown in Fig. 2.44 (Kobayashi et al. 2016). It is seen that wavelengths of the emission peaks are duly red-shifted compared with that of the absorption peak by ca. 100 nm (liquid phase) and by ca. 140 nm (solid phase) due to the Stokes shift as usual. The calculated absorption spectrum for this molecule in the same fashion with Fig. 2.39 is shown in Fig. 2.45. The origin of the calculated main peak

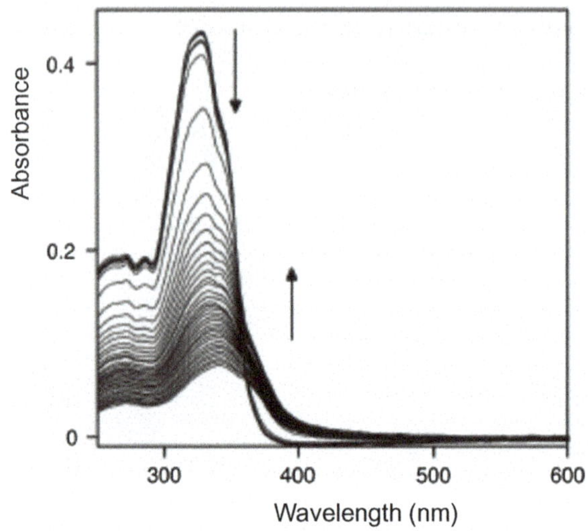

Fig. 2.41 Experimental UV-vis absorption spectra of the benzotrithiopehene derivative with decreasing temperature from 60 to 10 °C in steps of 1 °C in MCH solution of the concentration of 5.0×10^{-6} mol/l. Reprinted with permission from Ikeda et al. (2017). Copyright 2017 American Chemical Society

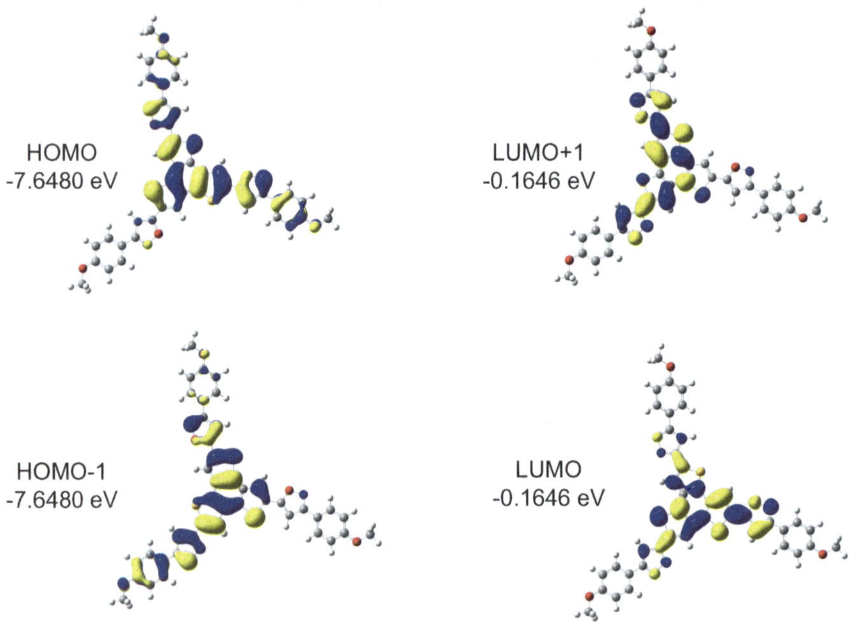

Fig. 2.42 MO patterns and their energies concerning the main peak of the UV-vis absorption of the benzotrithiophene derivative in vacuum (C_3 symmetry) obtained by DFT/ωB97X-D/6-31G**

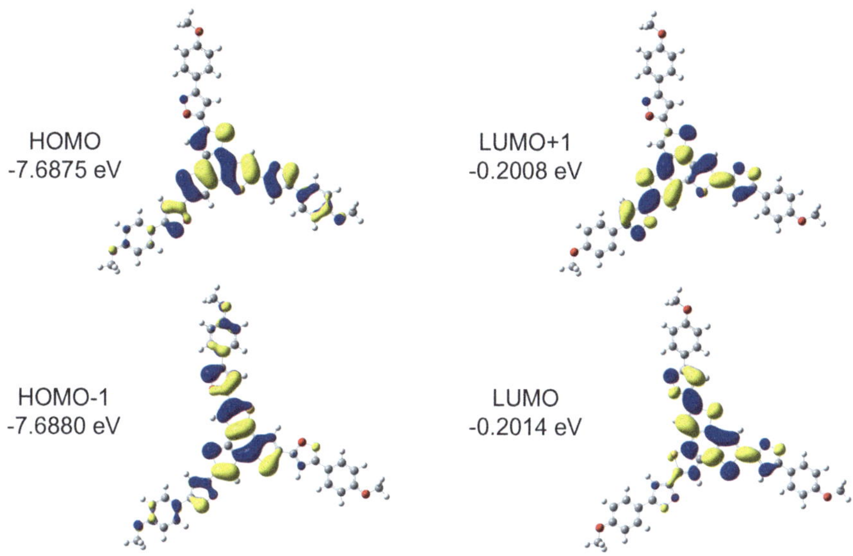

Fig. 2.43 MO patterns and their energies concerning the main peak of the UV-vis absorption of the benzotrithiophene derivative in MCH solution (C_1 symmetry) obtained by DFT/ωB97X-D/6-31G**

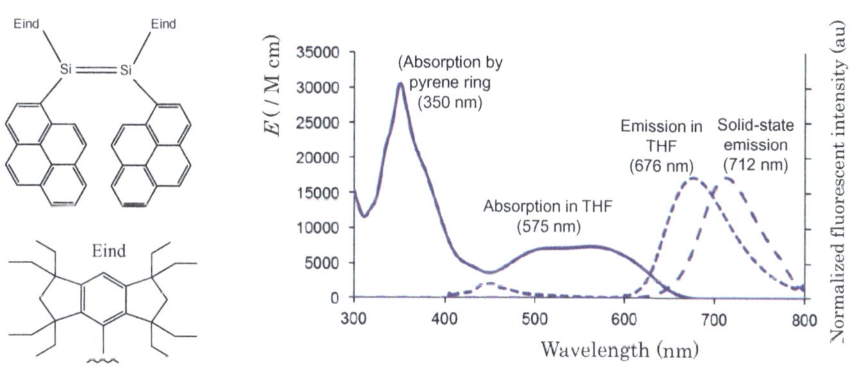

Fig. 2.44 Experimental UV vis absorption spectra (solid line) and emission spectra (dashed) of (*Z*)-1,2-di(1-pyrenyl)disilene containing bulky Eind (1,1,3,3,5,5,7,7-octaethyl-*s*-hydrindacen-4-yl) groups. Reprinted with permission from Kobayashi et al. (2016). Copyright 2016 American Chemical Society

at ca. 560 nm comes from the two transitions from the HOMO → LUMO and the HOMO → LUMO + 1 transitions as listed in Table 2.14. The concerning MO patterns are shown in Fig. 2.46 by which it is understood that these absorptions come from the intramolecular charge transfer from the π conjugation at Si = Si bond to the pirenyl rings. The absorption spectrum giving a peak at ca. 350 nm comes from those within the pyrene rings and omitted here.

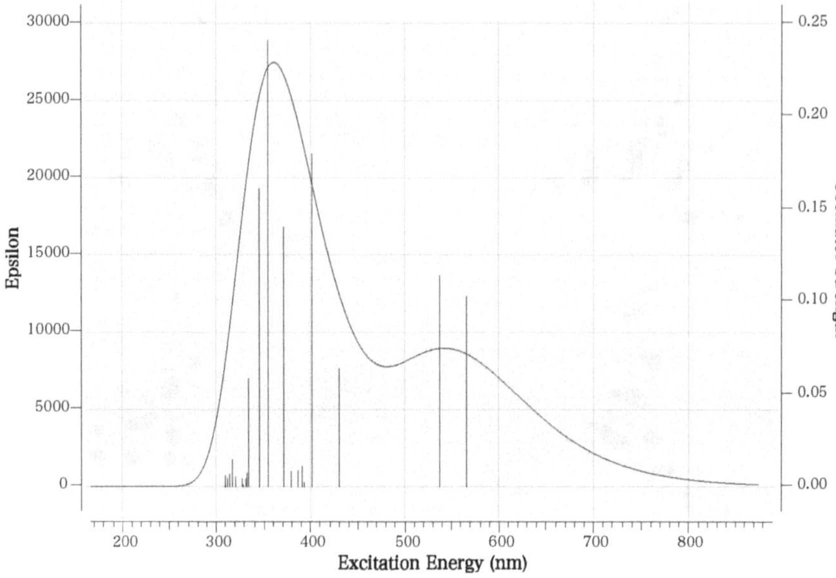

Fig. 2.45 Calculated UV-vis spectrum of (*Z*)-1,2-di(1-pyrenyl)disilene containing bulky Eind (1,1,3,3,5,5,7,7-octaethyl-*s*-hydrindacen-4-yl) groups in vacuum (C_1 symmetry) comparable with those in Fig. 2.44. The calculation was performed by TD-DFT/B3LYP/6-31G**. Reprinted with permission from Kobayashi et al. (2016). Copyright 2016 American Chemical Society

Table 2.14 Calculated data of the main optical absorption of the 1-pyrenyldisilene derivative with the Eind groups[a] in vacuum

Condition[b]	Excitated state	Absorption		
		Wavelength (nm)	Oscillator strength	Corresponding transition with the coefficient
$S_0 \to S_1$ at $\mathbf{R}(S_0)$	1	565.96	0.1025	HOMO → LUMO 0.6982
	2	537.76	0.1141	HOMO → LUMO + 1 0.6966
$S_0 \to S_1$ at $\mathbf{R}(S_1)$[c]	1	765.46	0.0716	HOMO → LUMO 0.7032
	2	617.37	0.1427	HOMO → LUMO + 1 −0.6910

[a]Eind = 1,1,3,3,5,5,7,7-octaethyl-*s*-hydrindacen-4-yl

[b]The S_0 and S_1 geometries ($R(S_0)$ and $R(S_1)$) were optimized by DFT/B3LYP/6-31G** and TD-DFT/B3LYP/6-31G**, respectively. The absorption data were obtained by the TD-DFT scheme from Kobayashi et al. (2016)

[c]The absorption wavelengths are regarded as those of the fluorescent emission in Fig. 2.49

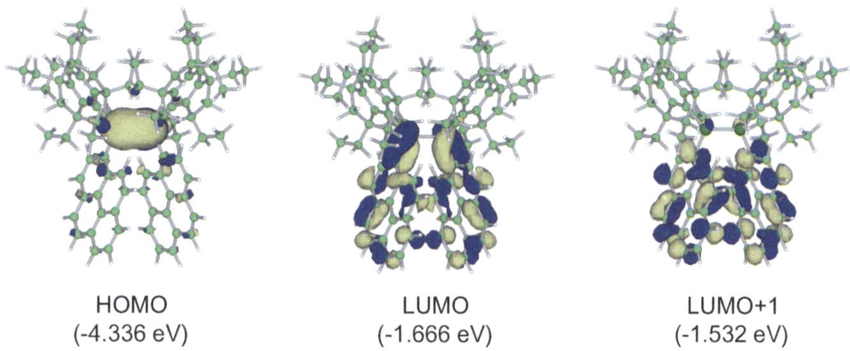

Fig. 2.46 The MO patterns and their energies concerning the main peaks of the UV-vis absorption spectrum of (*Z*)-1,2-di(1-pyrenyl)disilene containing bulky Eind (1,1,3,3,5,5,7,7-octaethyl-*s*-hydrindacen-4-yl) groups. The calculation was performed by DFT/B3LYP/6-31G**. Reprinted with permission from Kobayashi et al. (2016). Copyright 2016 American Chemical Society

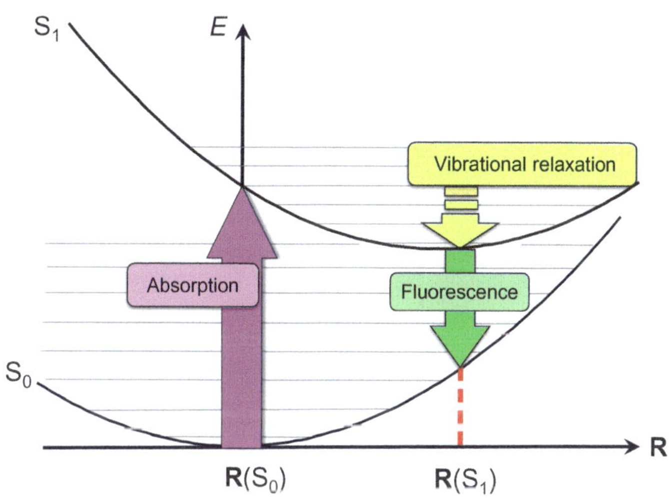

Fig. 2.47 Concept of emission from the excited state S_1 giving fluorescence

The stokes shift is understood by the illustration in Fig. 2.47 compared with the ordinary absorption peak based on the vertical transition. Actual calculation for the emission spectrum can be performed as the calculation of the absorption from the ground state with the molecular structure with the nuclear coordinates $\mathbf{R}(S_1)$ obtained by structural optimization at the S_1 state using the TD-DFT scheme. This strategy is shown in Fig. 2.48. The emission spectrum thus obtained is shown in Fig. 2.49 with the oscillator strength listed also in Table 2.14. Note that there are emission peaks at ca. 640 nm comparable with the experimental data of 676 nm observed in THF solution (see Fig. 2.44).

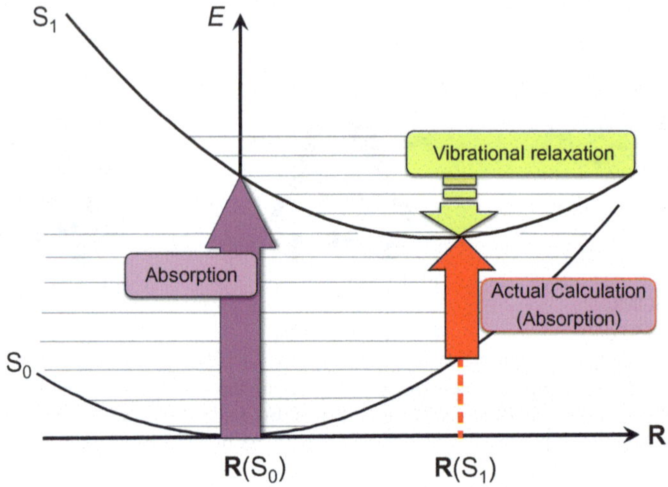

Fig. 2.48 The emission in Fig. 2.47 obtained by calculation of the absorption from the ground state with the structure $\mathbf{R}(S_1)$

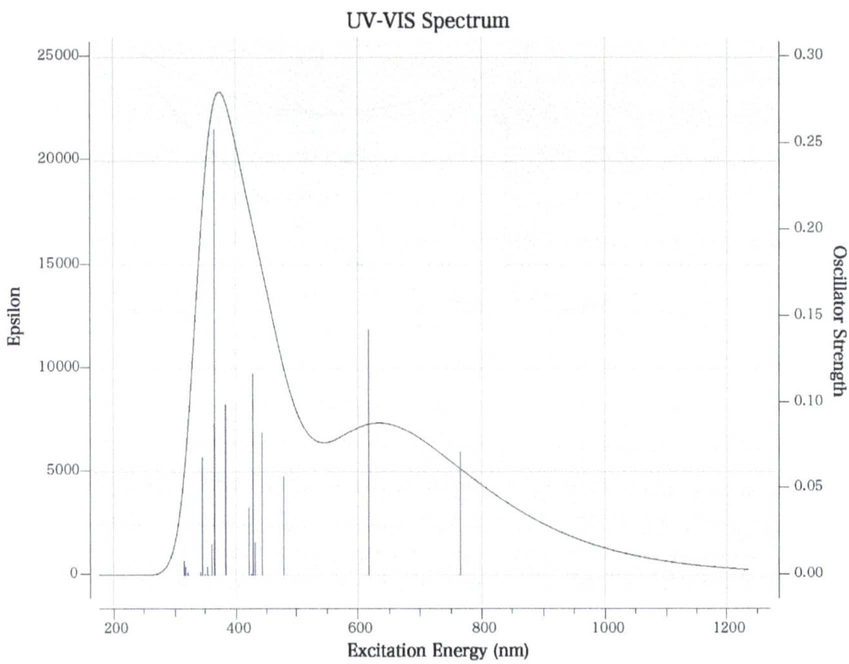

Fig. 2.49 Calculated emission spectrum of (*Z*)-1,2-di(1-pyrenyl)disilene containing bulky Eind (1,1,3,3,5,5,7,7-octaethyl-*s*-hydrindacen-4-yl) groups in vacuum (C_1 symmetry) comparable with those in Fig. 2.44. The calculation was performed by TD-DFT/B3LYP/6-31G** Reprinted with permission from Kobayashi et al. (2016). Copyright 2016 American Chemical Society

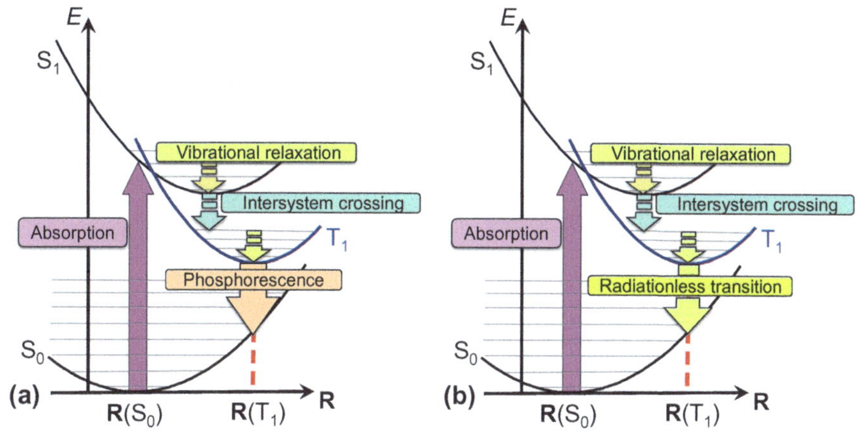

Fig. 2.50 Concept of transition from the excited state T_1 giving **a** phosphorescence and **b** radiationless transition

The phosphorescent emission with the transition from T_1 to S_0 becomes possible to the molecules containing, what is called, heavy elements with large spin-orbit coupling constants. The spin-orbit coupling allows mixing of the ground-singlet and the 1st triplet wavefunctions such as Ψ_{S_0} and Ψ_{T_1}, which effectuates the phosphorescent emission as illustrated in Fig. 2.50a, based on which the wavelength of the phosphorescence can theoretically be understood in a similar way to what is elucidated in Figs. 2.47 and 2.48. Small and insufficient coupling mainly causes radiationless transition such as eventual heat dissipation from T_1 to S_0 as shown in Fig. 2.50b instead of phosphorescent emission as in usual organic molecules not containing heavy elements.

For instance, in Table 2.15 are listed the experimental data of phosphorescent emission found in heterofluorene derivatives containing B, Al, Ga, or In atom in Fig. 2.51

Table 2.15 Data of the observed emission spectra of heterofluorene derivatives containing B, Al, Ga, or In measured at 77 K in 2MeTHF (1.0×10^{-4} M) and variables related to the spin-orbit coupling[a]

	B	Al	Ga	In
λ_{phos} (nm)[b]	446, 479	449, 481	463, 488	464, 489
λ_{fluo} (nm)[b]	314, 329	337, 351	334, 348	308, 324
Φ_{phos}/Φ_{total}	0.26	0.41	0.93	0.99
ζ (cm^{-1})[c]	10	62	464	1183
α ($\times 10^{-6}$)[d]	7.34	4.23	−64.2	91.1

[a]From Matsumoto et al. (2015)
[b]Excited at 282 nm
[c]From Montalti et al. (2006)
[d]See text

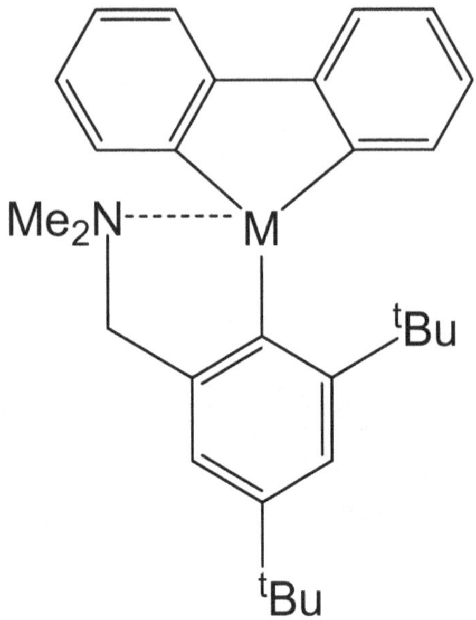

Fig. 2.51 Molecular structure of heterofluorene: M = B, Al, Ga, or In. Reproduced from Matsumoto et al. (2015) with permission from the Royal Society of Chemistry

with the calculated results of phosphorescence (Matsumoto et al. 2015). Theoretical calculations with the DFT/B3LYP/6-31G** for B-, Al-, and Ga-fluorenes, and with the DFT/B3LYP/LANL2DZ for Ga- and In-fluorenes were performed for the structural optimization at the S_0 state and TD-DFT/UB3LYP with the same bases at the T_1 state. The calculated phosphorescence for Ga- and In-fluorenes were at 523.13 and 525.87 nm, respectively, comparable with the experimental data in Table 2.15. In the description of phosphorescence, it is essential to estimate the total wavefunction of the T_1 state with mixing of the S_0-state wavefunction through the spin-orbit coupling interaction with the coefficient α as

$$\Phi_{\mathrm{Total}} = \Psi_{\mathrm{T_1}} + \alpha \Psi_{\mathrm{S_0}} \tag{2.42}$$

The calculated α is listed in Table 2.15 as well. Thus the phosphorescent behavior can also be estimated by theoretical calculation.

2.5.3 *Circular Dichroism (CD) Spectrum*

In organic chemistry field, it is well known that molecules having asymmetric carbon (or Si, Ge) is optically active. These optically active or chiral molecules cause optical spectroscopic effects such as optical rotation, optical rotatory dispersion, and circular dichroism (CD) by using circularly polarized light. These phenomena take place by difference between optical absorptions of chiral molecules obtained by left-hand polarized light and right-hand one, where the enantiomer shows the mirror image spectra. This behavior directly comes from rotatory strength (R) of the polarized light manifested by the electronic structure of a chiral molecule (Hansen and Bouman 1980; Helgaker and Jørgensen 1991). The rotatory strength (R) is theoretically calculated by the product of electric transition dipole moment and magnetic transition dipole moment as

$$R = \mathrm{Im}\left\{\left(\int \Psi_0^* \hat{\mu} \Psi_{\mathrm{S}_n} \mathrm{d}\tau\right)\left(\int \Psi_0^* \hat{m} \Psi_{\mathrm{S}_n} \mathrm{d}\tau\right)\right\} \tag{2.43}$$

where Im signifies the imaginary part, Ψ_0 and Ψ_{S_n} the wavefunctions, respectively, of the ground and the S_n excited states like in Eq. (2.40), and $\hat{\mu}$ the electric dipole operator and $\hat{m}$ the magnetic dipole operator defined as follows

$$\hat{\mu} \equiv -e \sum_i \mathbf{r}_i \tag{2.44}$$

$$\hat{m} \equiv -\frac{ie\hbar}{2mc} \sum_k (\mathbf{r}_k \times \nabla_k) \tag{2.45}$$

with the usual physical constants. The electric dipole operator $\hat{\mu}$ is essentially the same with that used in Eq. (2.40).

Analyses of the above optical effects by theoretical calculation and comparison with the experimentally obtained CD spectrum would be of use to decide the chirality of molecules or absolute configuration of molecules. Note that such plausibility by calculation normally requires more quantitative estimation of the excited state by the calculation using the post HF scheme in terms of elaborate configuration interaction (CI) methods such as MRCI, CC, or SAC-CI (see Sect. 3.1.2) whereas the conventional CIS calculation would not be sufficient.

For instance, comparison of the experimental CD spectrum of *trans*-(2*S*,3*S*) dimethyloxirane is shown in Fig. 2.52 with the calculated rotatory strengths (Carnell et al. 1994). In this calculation, the MRD-CI (multireference singles and doubles CI) method was employed with the large basis set of triple zeta quality augmented with d-polarization functions and diffuse functions (see Sect. 3.6). It would be of interest to point out that in this calculation the localized MO (LMO) was considered with nearly complete valence electrons. It is seen that the experimental and the calculated data are in plausible agreement.

In Figs. 2.53 and 2.54 are shown the comparison of the CD spectra with those obtained by experimental observation and by calculation of Z- and B-DNA's having the helical structures (Miyahara et al. 2013). The helical molecules or polymers are also optically active, where Z-DNA has a left-handed double helicity and B-DNA right-handed. In the calculations were employed the tetrameric oligomer models for each DNA to avoid calculation load. Each molecule was structure-optimized by the DFT/B3LYP/6-31G** method and the calculations of the excited states and of the rotatory strengths were performed by the SAC-CI method using the D95 basis set

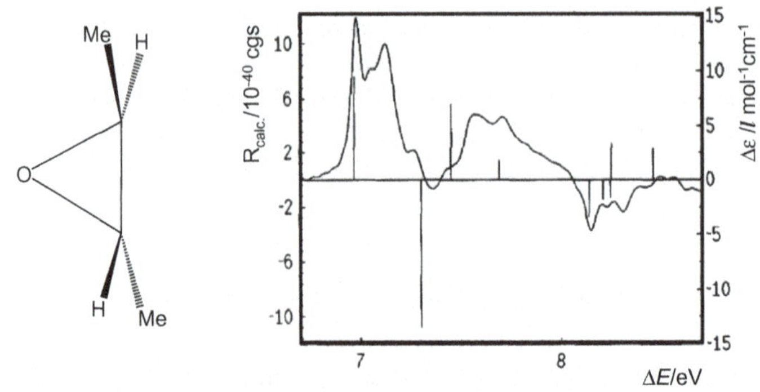

Fig. 2.52 The molecular structure of *trans*-(2*S*, 3*S*)-dimethyloxirane and the experimental CD spectrum with the calculated rotatory strengths represented by bar graphs. The calculated excitation energies have been shifted by 0.25 eV to fit the first transition energy at 6.97 eV. Reprinted from Carnell et al. (1994), Copyright 1994, with permission from Elsevier

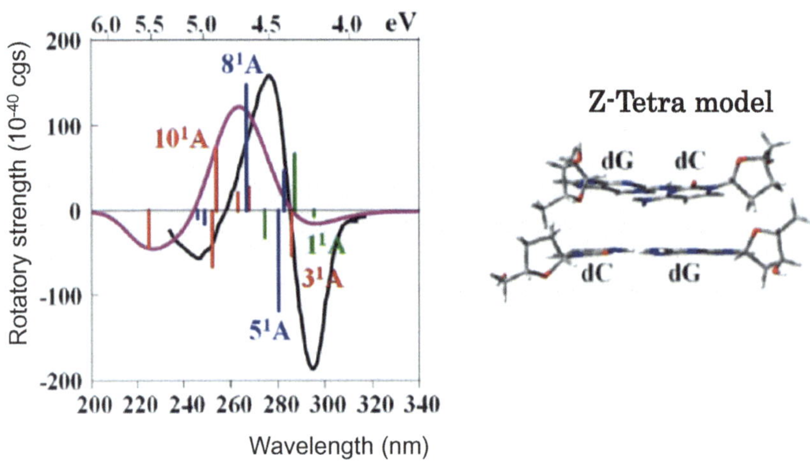

Fig. 2.53 CD spectra obtained by experiment (black) and by calculation (magenta) for Z-DNA (left-handed double helical structure). The latter spectrum was obtained by convoluting the calculated rotatory strengths indicated by bar graphs. For the calculation a tetramer model in the right side was employed. The calculated excitation energies have been shifted by 0.5 eV to the lower values. Reprinted with permission from Miyahara et al. (2013). Copyright 2013 American Chemical Society

without polarization functions, and the energies of the active orbitals were chosen between −1.1 and +1.1 hartree. These calculation results are in good agreement with the experimental data.

It is further pointed out that the vibrational CD (VCD) in the infrared (IR) region is often utilized for augmentation of the preciseness of the absolute configuration of optically active molecules. Theoretical background on VCD has been developed (Nafie and Freedman 1983) and its calculation routine is implemented into

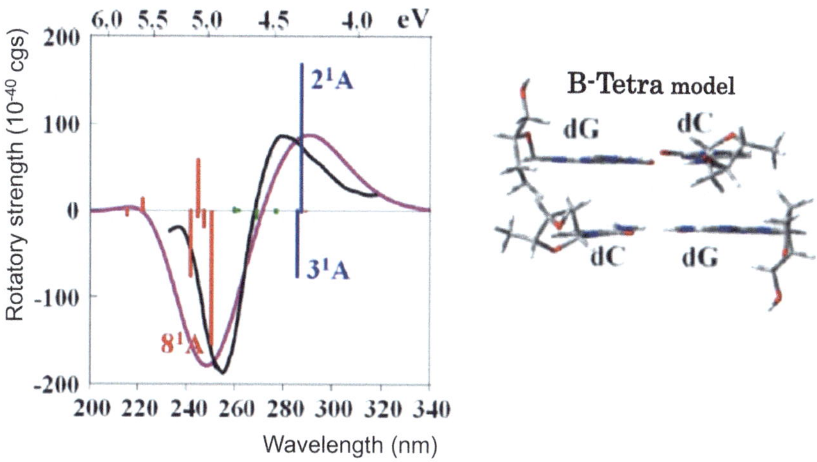

Fig. 2.54 CD spectra obtained by experiment (black) and by calculation (magenta) for B-DNA (right-handed double helical structure). The latter spectrum was obtained by convoluting the calculated rotatory strengths indicated by bar graphs. For the calculation a tetramer model in the right side was employed. The calculated excitation energies have been shifted by 0.5 eV to the lower values. Reprinted with permission from Miyahara et al. (2013). Copyright 2013 American Chemical Society

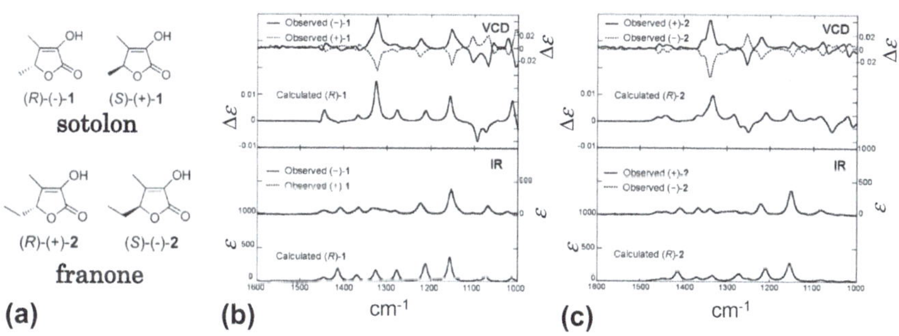

Fig. 2.55 **a** Absolute configurations of sotolon (1) and maple furanone (2) and the observation and calculation data of VCD and IR spectra with respect to **b** sotolon (1) and **c** maple furanone (2). See text for the detailed calculations. Reprinted with permission from Nakahashiet al. (2011). Copyright 2011 American Chemical Society

the conventional MO calculations software. It is mentioned here that the comparison of the observed and the calculated VCD spectra have been accumulated as in Fig. 2.55 (Stephens et al. 2007; Nakahashi et al. 2011).

2.6 Mechanical Properties

Theoretical-chemistry calculation also works as a kind of valuable probe for the mechanical phenomena occurring in molecules. Furthermore, theoretical analyses of quantities related to elastic constant of linear polymers are of use toward essential understanding of the response of 1D-polymers to the stress caused by the external force field such as tensile strength along the main-chain axis. In this section, some examples of application to mechanical properties of molecules and regular 1D polymers are to be afforded for comprehension of efficiency of theoretical calculation on this theme.

2.6.1 *Strains in Molecules*

Some molecules can store strain energies inside them and they often show higher chemical reactivity or spontaneous isomerization to release strain energies compared with less strained ones. There can be a couple of patterns to store molecular strains: (1) bond-angle strain, (2) steric strain, and (3) torsional strain, each of which is connected to the peculiar molecular structure. The bond-angle strain causes the deviation from the normal hybridization angles, that is, 109.5° for sp^3, 120° for sp^2, and 180° for *sp* hybridizations. The steric strain comes from "too close" atoms in molecule causing severe interelectron repulsions and, hence, is also referred to as steric repulsions. The torsional strain can happen in association with eclipsed or gauche interactions. All of these strains can be estimated by theoretical calculations on the total energy or heat of formation of the concerning molecule.

In this subsection, theoretical calculation dealing with the bond-angle strain taking as the example of interest is to be described. Small-ring molecules with three or four membered rings usually possess high strains inside them or, in other words, they automatically store mechanical energy in themselves. This is because the bond angles largely change from the normal bond angles such as those for sp^3 (109.5°) and sp^2 (120°) hybridizations. For instance, cyclopropane and cyclobutane have the bond angles of 60° and 90°, respectively, being considerably smaller than 109.5°, that is, they are out of the normal sp^3 hybridization. This situation causes rather peculiar electronic structures compared with those of the strain-free molecules. For instance, in Fig. 2.56a there is seen an electron-deficient hollow inside the triangle formed by the three C–C bonds with the electron density rather tending to direct outside the molecular skeleton of cyclopropane. On the other hand, such behavior is not seen in the normal propane molecule.

More quantitative way of describing magnitude of the strain is to examine the strain energy based on the MO calculation scheme in an adequate manner. There can be several ways of definition of strain energies in theoretical chemistry, and perhaps the most popular one would employ the concept of homodesmotic reaction. This belongs to isodesmic reactions, in which the total numbers of each bond type

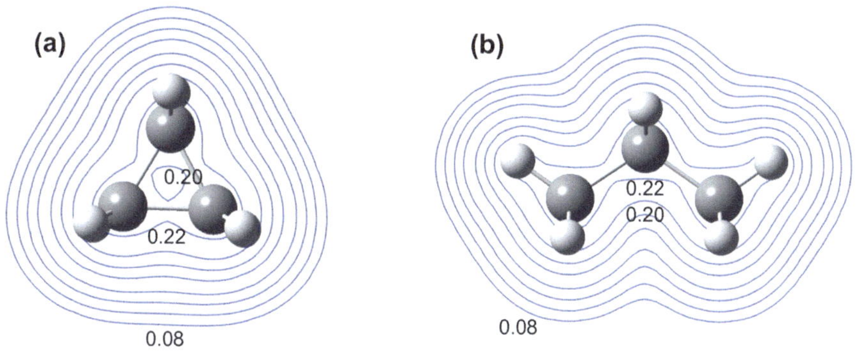

Fig. 2.56 Electron density contours on the three C planes of **a** cyclopropane and **b** propane, obtained by DFT/B3LYP/6-31G. The outermost density contour indicates 0.08 e/au^3

(single, double, or triple bond) in the reactants and in the products are the same. The homodesmotic reaction further requires the following conditions (George et al. 1976):

(1) There are equal numbers of C atoms with each hybridization type in the reactants and the products, and
(2) There are equal numbers of atoms such as C or Si with zero, one, two, and three H atoms attached in the reactants and the products.

Typical examples of homodesmotic reaction, for instance, are shown in Fig. 2.57. It is seen that in Fig. 2.57a five sp^3–C atoms and four sp^2–C atoms are present in the both reactants and products. Also, there are two C=C bonds and four C–C bonds. In Fig. 2.57b eight sp^3–C atoms and four sp^2–C atoms are present, and two C=C bonds and six C–C bonds. Hence these two reactions are considered to be homodesmotic. Both of these reactions should be exothermic since cyclopropane and cyclobutane naturally release the strain energy upon the reactions.

In Table 2.16 are listed examples of the strain energies obtained by the combination of theoretically estimated heat of formation ΔH_f for each molecule involved in the concerning homodesmotic reaction. The usual thermochemical routine in the

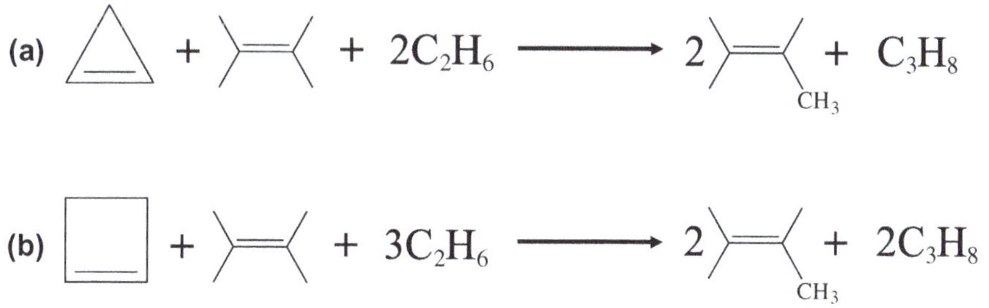

Fig. 2.57 Examples of homodesmotic reactions

Table 2.16 Examples of strain energies (in kcal/mol) of several molecules obtained by theoretical calculation (DFT/B3LYP/6-311++G(3df, 2p) //B3LYP/6-31G*) and the experimental data

Molecule	Theor. calc.[1]	Exp. data	
		Homodes-motic[2]	Methylene fragment[3]
Cyclopropane 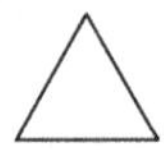	25.5 (35.5)[4]	30.6	27.5
Cyclobutane 	22.6 (12.9)	30.3	26.5
Cyclopentane 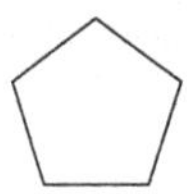	4.7 (3.0)	4.4	6.2
Cyclopropene 	55.5 (34.5)[4]	54.5	55.2
Cyclobutene 	28.7 (9.1)	33.5	28.4
Cyclopentene 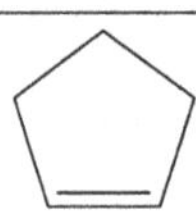	4.5 (0.9)	5.5	4.1
[2.2.2]Propellane 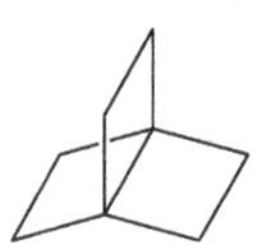	79.2 (43.0)	-	89
Spirobicyclobutane 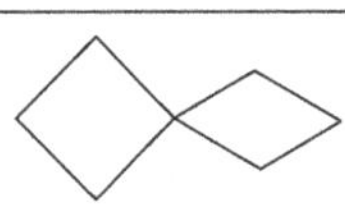	41.5 (23.8)	-	-

[1] In parentheses the values of molecules changing all carbon to silicon atoms are listed From Naruse et al. (2006) except for that marked with superscript 4
[2] Based on the homodesmotic reactions (see text and Fig. 2.57). From George et al. (1976)
[3] Compared with the strain-free methylene fragments. From Wiberg (1986)
[4] From Iwamoto et al. (2000)

MO calculation software is applicable to enumeration of the ΔH_f values (see, e.g., Sect. 2.1). There are two ways of experimental estimations of the strain energy with the use of the observed ΔH_f data (Cox and Pilcher 1970) as follows: (i) utilization of the homodesmotic reactions as employed in the above (George et al. 1976) and (ii) comparison of the observed ΔH_f value of the concerning molecule with that of assembly of the strain-free standard molecule fragments separately considered (Wiberg 1986). In the process (ii), for instance, the strain energy of cyclopropane is obtained as the difference of ΔH_f values of that and the strain-free three methylene fragments derived from ΔH_f of cyclohexane selected as the standard, although there can be a certain arbitrariness in selection of the standard molecule as a matter of course. These experimental-based strain energies are also listed in Table 2.16.

It is seen that the strain energies theoretically obtained are in reasonable agreement with those based on the experimental estimations. The more distorted bond angles duly cause the more strain energies, which is more remarkable in the molecules with the unsaturated bonds. It is noted that substitution of all the carbon to silicon atoms decreases strain energies as a whole except for silacyclopropane. Moreover, introduction of unsaturated bond in Si-substituted molecules has been shown to bring about less strain energies, which has been explained by the strain-energy relaxation associated with the presence of high π-orbital energy and low σ*-orbital energy of the Si–H bonds through the orbital interaction from the theoretical viewpoints (Naruse et al. 2006).

It is noted that the experimental ΔH_f is generally derived from the observed heat of combustion of each molecule but this quantity is sometimes difficult to be obtained when the molecule can only exist under special conditions such as impregnation in the matrix of other materials and so on. Hence in such case or for yet unsynthesized molecules only the theoretical calculation of ΔH_f would become accessible.

2.6.2 Elastic Constant of Polymers

One of the conspicuous and crucial mechanical properties manifesting in the linear oligomeric or polymeric chains would be the elastic constant. This property is generally represented by Young's modulus which decides the strain replying to compression or tensile strength along the main chain of the polymer. The strain generally appears as response of the system to the external force causing stress to the material. Although these quantities are rather macroscopic variables in macromolecules or their oligomeric model structures, analysis of those from microscopic viewpoints using theoretical chemistry is also possible as described in this subsection.

There are several experimental ways for measurement of elastic constant of pure crystal regions of polymers such as X-ray diffraction method, inelastic neutron scattering, and Raman scattering as well as macroscopic measurement of strain curve (Tashiro 1993). Various theoretical estimations of elastic constant of polymers have also been performed with the use of conventional molecular mechanics (Špitalský and Bleha 2001), force field method derived from HF/6-31G**/MP2 calculation

(Dasgupta et al. 1996), ab initio molecular dynamics (MD) method (Hageman et al. 1997), and CO (see Sect. 3.3) methods using Hessian matrix method (Bartha et al. 2000), and so on.

In the following two subsections, theoretical estimations of the quantities related to elastic constants of several kinds of oligomers based on the HF method, and those of polymers with infinite chain length based on the direct estimation using the CO method are to be described. Each method has its own feature but has common aspect in directly dealing with deformation modes of the structural parameters such as bond lengths, bond angles, and dihedral angles due to the elongation of the original polymers along the longitudinal direction.

2.6.3 *Young's Moduli of Oligomeric Species*

Here is discussed the response to mechanical stress by compression or tensile strength applied to the linear oligomeric chains. In this subsection, the oligomeric materials are to be called polymers for simplicity, although the MO method is effective only to oligomers with finite length. In this sense, the obtained result of Young's modulus in the below ought to be considered with a certain reservation and/or extrapolated to that of the infinite polymer chain.

Under small mechanical stress σ caused by the above strength along the main-chain direction of a polymer in general, the response appears as strain ε and is represented as

$$\sigma = E_l \varepsilon \tag{2.46}$$

within the range of linear relationship. The coefficient E_l denotes Young's modulus with respect to the polymer deformation. Since the stress σ can further be expressed as F/S with F and S being, respectively, the force applied to the polymer and its cross-sectional area, E_l can be expressed as

$$E_l = \frac{F/S}{\varepsilon} \tag{2.47}$$

This equation implies that E_l has the same dimension as that of pressure such as dyn/m^2 or N/m^2, since strain is a dimensionless quantity. If we consider stretch of a polymer along its main-chain direction, for instance, ε can be expressed as $\Delta L/L$ (see Fig. 2.58). Hence Young's modulus E_l can also be written as

$$E_l = \frac{F/S}{\Delta L/L} \tag{2.48}$$

In the theoretical calculation, one can estimate destabilization energy ΔW caused by stretch ΔL of the concerning polymer from its original length L at the optimized

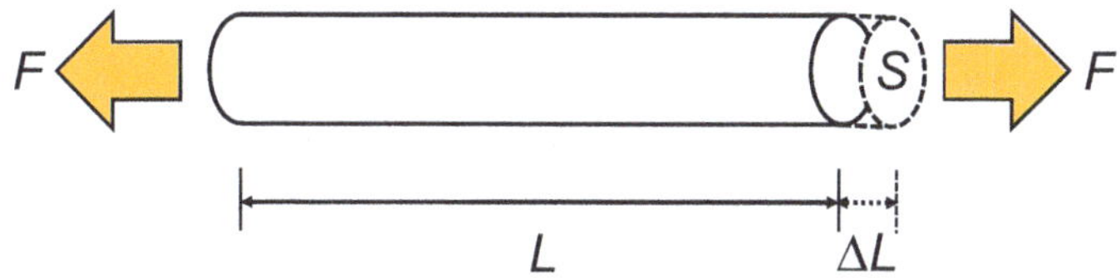

Fig. 2.58 Schematic drawing of tensile strength along the main-chain direction of a polymer

structure. The relationship

$$\Delta W = F \Delta L \tag{2.49}$$

then allows to obtain E_l without using F, which gives the expression

$$E_l = \frac{F/S}{\Delta L/L} = \frac{\Delta W/S}{\Delta L^2/L} = \frac{\Delta W}{\Delta V(\Delta L/L)} \tag{2.50}$$

where ΔV stands for the volume change (=$S\Delta L$) of the polymer under the strain.

In Table 2.17 are listed the calculated data of E_l of the models for polyyne and polystaffane having finite chain lengths (see Fig. 2.59) obtained by the quantum chemical calculation with two kinds of compression concepts in mechanical engineering, that is, buckling and bending mode employed for estimation of their radii (Itzhaki et al. 2005). Skeleton of polystaffane could be regarded to model one-dimensional diamond. Also in Table 2.17 are listed the data of models for single-walled carbon nanotubes (SWCNT) with different kinds of chiral indices (n, m) having endcaps, graphene (van Lier et al. 2000), and polyamide-6 (nylon6) (Peeters et al. 2003) in Fig. 2.60 with their radii estimated by various methods.

From the numerical values for polymer models in Table 2.17, it is understood that polyyne (or carbine) is likely to have large Young's modulus of more than one order of magnitude compared with diamond due to existence of C≡C bonds, although that material hardly exists in actuality because of unstable and highly reactive characteristics under ambient condition. This quantity for graphene model is comparable to the experimental data of diamond and SWCNT models. Moreover, it seems that polymers whose main chain consists of C–C bonds have less Young's moduli by one order of magnitude as a whole. For instance, polyamide-6 (nylon6) model has E_l of about 300 GPa being still larger than that of stainless steel (194–199 GPa) (Rumble 2018).

2.6.4 f-*Values of Polymers with Infinite Chain Length*

The optimized structures of polymers with infinite chain length or, in other words, those having infinite repetition of the unit cell can theoretically be obtained by the CO method (see Sect. 3.3). Hence the elastic constant of polymers can also be

Table 2.17 Young's moduli estimated by theoretical calculations

Material	Radius r (in Å) or cross-sectional area (in Å^2) (estimation method)	Young's modulus E_l (in GPa)[a]
Polyyne[b,c]	Radius: 0.35 ± 0.005 (Buckling method)	40000 ± 2000
	Radius: 0.33 ± 0.02 (Bending-mode method)	47000 ± 6000
Staffane[b,d]	Radius: 0.76 ± 0.02 (Buckling method)	5300 ± 300 1063 (diamond exp. data[e])
SWCNT (9,0)[f]	Radius: 3.56 (Opt. CNT radius + van der Waals radius (1.7 Å))	1140
SWCNT (8,2)[f]	Radius: 3.65 (Opt. CNT radius + van der Waals radius (1.7 Å))	1030
SWCNT (5,5)[f]	Radius: 3.44 (Opt. CNT radius + van der Waals radius (1.7 Å))	1060
Graphene[f]	Area obtd. by: multiplication of thickness (twice of the van der Waals radius) and width (distance between the outer two anchor atoms)	1110
Polyamide-6 (α-form)[g]	Area: 18.02 (Cryst. struct.)	334

[a]Note that 10^9 Nm^{-2} = GPa
[b]Obtained by DFT/B3LYP/6-31G from Itzhaki et al. (2005)
[c]Average for oligoynes HC_mH (m = 4–20)
[d]Average for [3] and [4]staffanes
[e]From Huntington (1958)
[f]Closed CNT with endcaps and obtained by HF/6-31G** with elongation $\Delta L = L/100$ (L the original total length of the oligomer). From van Lier et al. (2000)
[g]Obtained by HF/6-31G** with elongation $\Delta L = L/100$ (L the original total length of the oligomer). From Peeters et al. (2003)

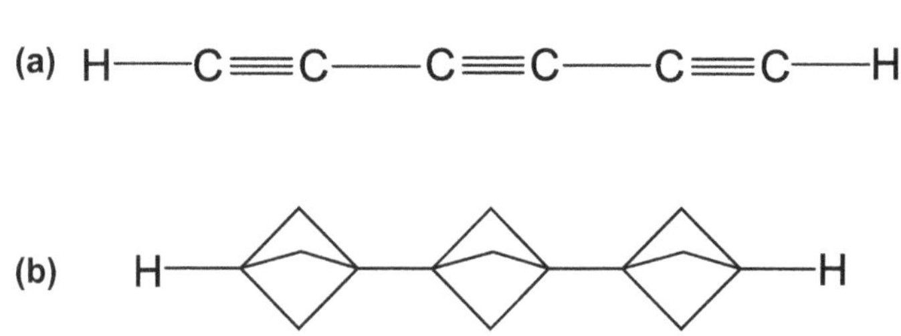

Fig. 2.59 Structures of **a** polyyne (HC_6H) and **b** [3]staffane

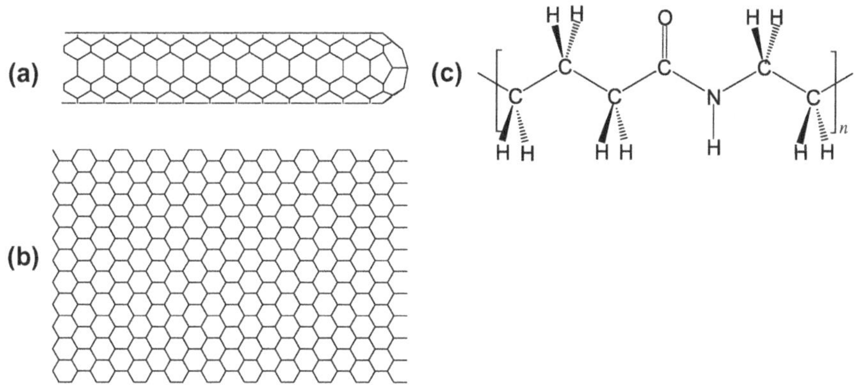

Fig. 2.60 Model structures of **a** SWCNT, **b** graphene, and **c** polyamide-6 ($n = 4$–16)

estimated in principle. In some polymers with rather long side chains, however, it may not be appropriate to consider the apparent cross-sectional area of those in obtaining Young's modulus, since the mechanical properties usually come from the response of the main chains. Hence, under such circumstances, it might be helpful to consider what is called f-value expressing the force required to give definite strain of the polymer chain along the longitudinal direction rather than Young's modulus itself. As a matter of fact, in Table 2.18, it is seen that the experimental f-values of poly(3-n-butylthiophene) and poly(3-n-hexylthiophene) are almost the same in spite of substantial difference between their Young's moduli due to different cross-sectional areas.

Expression of the f-value is more straightforward than Young's modulus and for the polymers with the infinite chain length is defined by

$$f = \frac{\Delta W}{\Delta L} = \frac{\Delta W/N}{\Delta L/N} \tag{2.51}$$

Table 2.18 Elastic constants of polymers obtained by DFT/B3LYP/6-31G** and by the experimental data[a]

Polymer	f-value (in 10^{-5} dyn)		Cross-sectional area (in Å^2)	E_l (in GPa)
	Calc.	Obs.	Obs.	Obs.
Polyethylene	2.59	4.28[b]	18.2	235
Polythiophene	5.46	4.79[c]	47.9	100
		4.78[d]	65.3	73
Polyselenophene	7.71	n/a	n/a	n/a

[a]From Nishino et al. (2014) (Calc. data have been further refined here)
[b]Exp. values from Nakamae et al. (1991)
[c]Exp. values for head-to-tail poly(3-n-butylthiophene)
[d]Exp. values for head-to-tail poly(3-n-hexylthiophene)

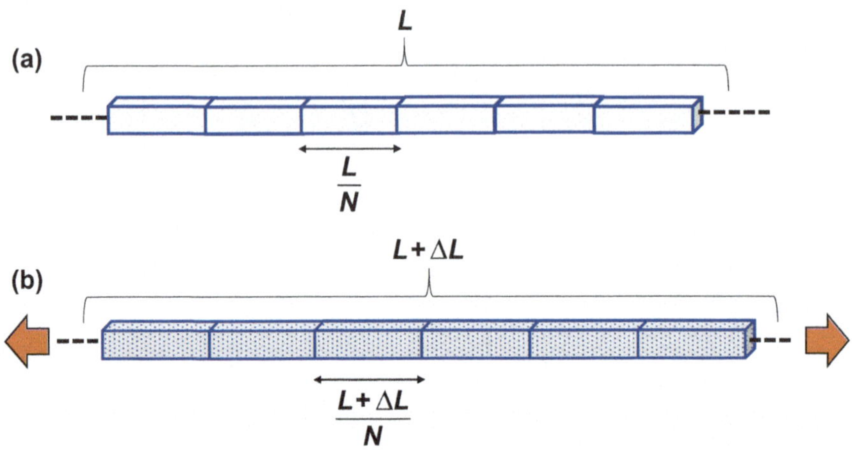

Fig. 2.61 Schematic representation of elongation stress to a polymer. Each rectangular denotes the unit cell. **a** Original polymer with the total length L and **b** elongated polymer by ΔL with the total length $L + \Delta L$. The number of the unit cell is indicated by N, which is actually infinite

where N designates the number of the unit cells formally being infinite. Hence $\Delta W/N$ and $\Delta L/N$ stand for the energy change and elongation of the polymer length, respectively, the both quantities being in the sense of per unit cell (see Fig. 2.61). The number N is formally infinity but can be reduced to the fraction with size of the unit cell as in Eq. (2.51). The quantity $\Delta L/N$ simply implies the elongation length per unit cell and, in this subsection, is taken as $0.01L/N$, that is, 1% elongation of the original length is considered as in the usual estimation of elastic constant. Also note that $\Delta W/N$ is the change in the energy per unit cell upon the elongation. The prescription to obtain the f-values are in what follows:

(1) The polymer with infinite chain length is first energetically optimized, from which the corresponding set of all the atomic coordinates in the unit cell and the unit-cell length L/N are obtained (see Fig. 2.61a). The most stable energy per unit cell represented by W/N is thus decided ($W < 0$).
(2) 1% elongation of the whole polymer length L is introduced as the strain along the polymer chain due to tensile strength. This is realized by making change of the optimized unit-cell length L/N obtained in (i) into $(L + \Delta L)/N$ with $\Delta L = 0.01L$. The resulting destabilized energy $\Delta W/N$ ($\Delta W > 0$) affords the data of f-values with the use of Eq. (2.51).

In Table 2.18 are shown the calculated results of f-values with respect to three kinds of polymers as shown in Fig. 2.62 along with the experimental data (Nishino et al. 2014). Note that polyselenophene has not yet been synthesized but that it can be treated in the theoretical calculation. It is seen that the f-values obtained by theoretical calculation are in reasonable agreement as a whole with the experimental values in the order of size. Accompanied change in structural parameters such as bond lengths and angles of these polymers upon the elongation are listed in Tables 2.19, 2.20 and 2.21

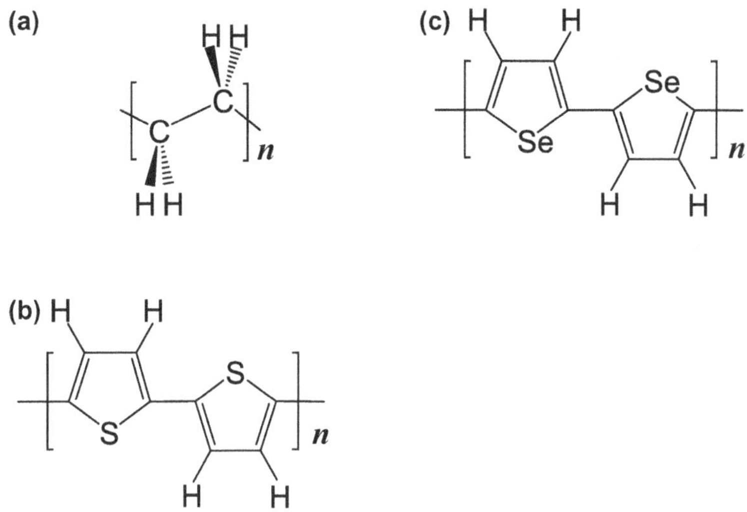

Fig. 2.62 Polymers considered: **a** polyethylene, **b** polythiophene, and **c** polyselenophene

Table 2.19 Change in structural parameters and bond orders of polyethylene upon the 1% elongation

	Bond lengths and angles (in Å and degrees, respectively)		Bond orders (Mulliken's bond population)	
	Before elongation	After elongation	Before elongation	After elongation
C^1C^2	1.5350	1.5443	0.738	0.736
$\angle C^1C^2C^3$	113.62	114.31	–	–
Unit-cell length	2.5692	2.5949	–	–

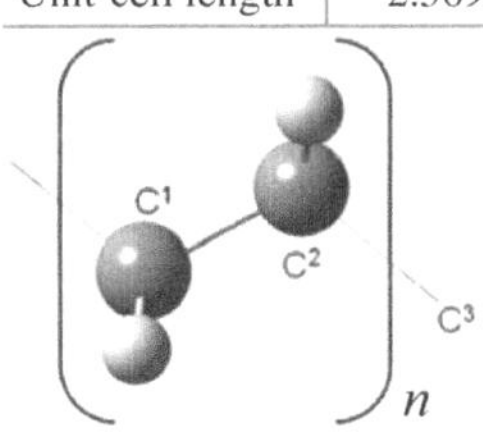

together with change in the bond orders. It is seen that bond lengths are lengthened and the bond angles made opened in all the polymers due to the elongation of the polymer chains but that the change in the bond orders reveals some characteristic behaviors. That is, although the bond orders are duly weakened in polyethylene, some of those in polythiophene and polyselenophene become a bit larger at pentagonal thiophene and selenophene rings. These peculiar behaviors of bond orders can only be found by theoretical analyses including the electronic structures, which might be helpful to understand mechanical properties of molecules in deeper sense.

Table 2.20 Change in structural parameters and bond orders of polythiophene upon the 1% elongation

	Bond lengths and angles (in Å and degrees, respectively)		Bond orders (Mulliken's bond population)	
	Before elongation	After elongation	Before elongation	After elongation
SC^4	1.7579	1.7670	0.634	0.628
$\angle C^1SC^4$	92.21	92.79	–	–
C^2C^3	1.4136	1.4166	0.934	0.936
C^3C^4	1.3821	1.3828	1.072	1.074
C^4C^5	1.4429	1.4540	0.690	0.692
$\angle C^2C^3C^4$	113.90	114.40	–	–
$\angle C^3C^4C^5$	129.23	129.23	–	–
$\angle SC^4C^5$	120.78	121.56	–	–
Unit-cell length	7.8505	7.9290	–	–

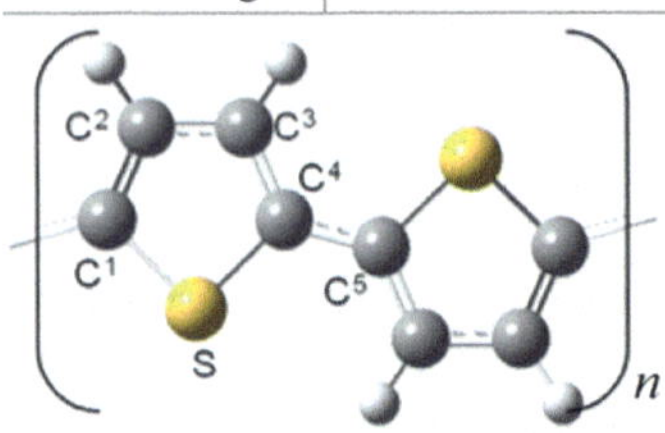

2.7 Chemical Reactions

Research of chemical reactions is one of the most important themes of theoretical chemistry in the both cases of actual or designed reactions. It had been indeed the long dream of chemists to afford interpretation and analysis of ungraspable chemical reactions which may even not be apt to take place. In this section, two major subjects thereof, orbital interaction and reaction path analyses, are to be described.

2.7.1 Orbital-Interaction Approach

There have been developed quite a few theoretical approaches to interpretation and prediction toward chemical reactions almost right after the commencement of quantum chemistry. Among others, the concepts of orbital-interaction based on the frontier orbital theory and orbital symmetry are rather simple but quite useful in that it can visualize the chemical reaction modes in terms of the MO pattern and symmetry (Fukui et al. 1952; Fukui 1971; Woodward and Hoffmann 1965, 1969). This approach utilizes the MO's of isolated molecules before the chemical reaction takes place. Chemical reactions are often accompanied by formation of the new bond(s) based on accumulation of electron density in the new-bond region between

Table 2.21 Change in structural parameters and bond orders of polyselenophene upon the 1% elongation

	Bond lengths and angles (in Å and degrees, respectively)		Bond orders (Mulliken's bond population)	
	Before elongation	After elongation	Before elongation	After elongation
SeC^4	1.8983	1.9089	0.686	0.676
$\angle C^1SeC^4$	88.11	88.66	–	–
C^2C^3	1.4143	1.4180	0.946	0.956
C^3C^4	1.3799	1.3809	1.050	1.070
C^4C^5	1.4369	1.4471	0.758	0.742
$\angle C^2C^3C^4$	116.36	116.91	–	–
$\angle C^3C^4C^5$	128.86	128.95	–	–
$\angle SeC^4C^5$	121.56	122.29	–	–
Unit-cell length	8.0856	8.1664	–	–

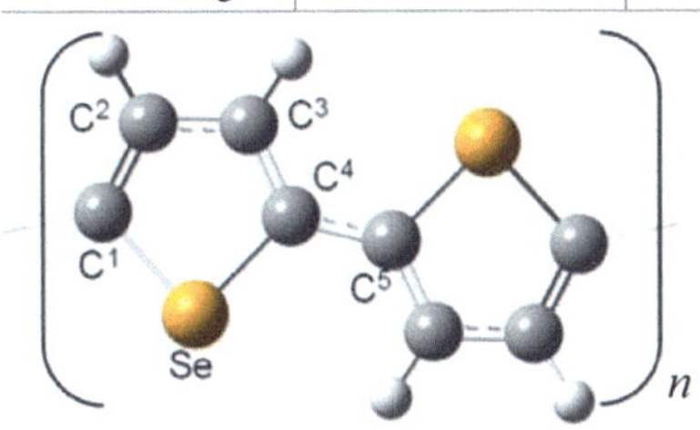

molecules A and B and/or by cleavage of a certain old bond(s) in those molecules due to diminution of electron density at the old bond region concerned.

For the accumulation mentioned above, it is natural to consider that the electron transfer occurs from A to B, in which the HOMO of A and the LUMO of B normally play important roles. That is, the molecule A acts as electron donor and B as acceptor extending the ordinary viewpoint of charge-transfer complexes. This kind of electron transfer relating with formation of new bond is often referred to as electron delocalization as well. In Fig. 2.63 is shown the schematic electron transfer from the HOMO of A to the LUMO of B normally causing electron accumulation between A and B so as to make a new bond therein resulting in the product molecule C. On the other hand, it is noted that, in terms of this electron delocalization, the original bonding characteristics in the HOMO is simultaneously weakened whereas the antibonding one in the LUMO is strengthened.

Let us consider here energetical stabilization $\Delta E_{r,s}$ due to the HOMO-LUMO interaction at the specific sites r and s of molecules A and B, respectively, as in Fig. 2.64 employing the simple perturbation theory which affords

$$\Delta E_{r,s} \propto -\frac{S_{rs}c^2_{r,\mathrm{HO(A)}}c^2_{s,\mathrm{LU(B)}}}{\Delta\varepsilon_{\mathrm{HO\,-\,LU}}} \tag{2.52}$$

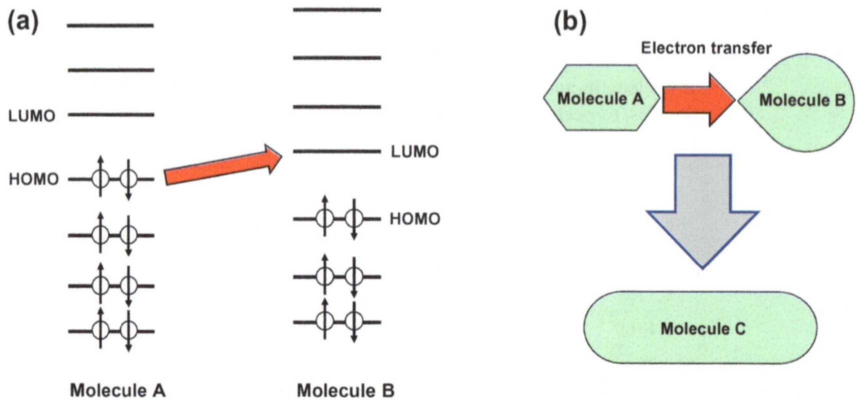

Fig. 2.63 **a** HOMO-LUMO interaction between molecules A and B resulting in electron transfer from the HOMO of A to the LUMO of B, and **b** schematic representation of formation of a new molecule C

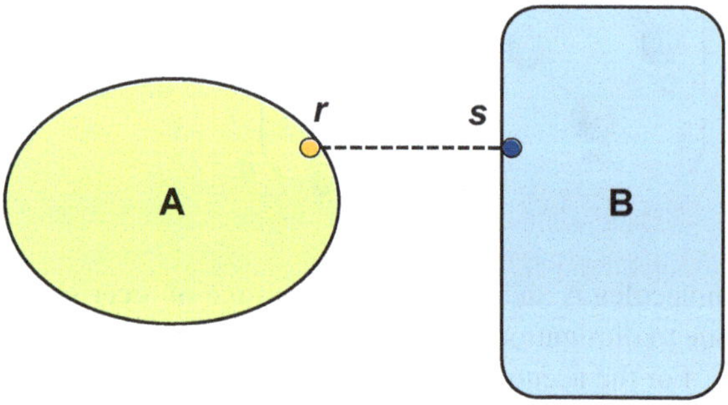

Fig. 2.64 Schematic representation of the HOMO-LUMO interaction between the sites r of A and s of B

where S_{rs} stands for the overlap integral of the AO's at the sites r and s, c_r the coefficient of the AO at r in the HOMO of A, c_s the coefficient of the AO at s in the LUMO of B, and $\Delta\varepsilon_{\text{HO-LU}}$ defined by $\varepsilon_{\text{LU(B)}} - \varepsilon_{\text{HO(A)}}$ (normally > 0) the energy difference of the HOMO of A and the LUMO of B. This energetical stabilization promotes favorableness of the initial stage of the reaction and possibly lowers the transition state of the reaction as illustrated in Fig. 2.65, when the following conditions are satisfied:

I. S_{rs} is positive and large,
II. $\Delta\varepsilon_{\text{HO-LU}}$ is small,
and
III. $c_{r,\text{HO(A)}}$ and $c_{s,\text{LU(B)}}$ are large.

Orbital-interaction approach basically utilizes the above principle based on perturbation theory and often affords prediction as well as understanding of the chemical reaction mode such as stereospecificity or stereoselection influencing the molecular structures of the reaction products. A couple of examples of orbital-interaction

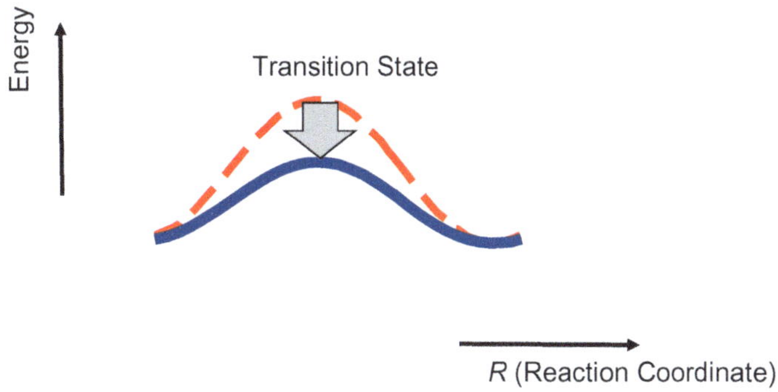

Fig. 2.65 Energy profile of chemical reaction along the reaction path expressed by red broken curves. Blue curves represent energetically more favorable reaction path with lower transition-state energy guaranteed by orbital interaction

approaches are shown below. Incidentally, it is noted here that for orbital-interaction analysis it is sometimes more convenient to use simple basis set encountered in the extended Hückel or semiempirical HF calculations (see Sect. 3.1), since they tend to afford clearer picture compared with rather complex basis set including diffuse orbitals or so suitable to consideration of quantitative energetics.

Note that, in consideration of the orbital interactions, we had better distinguish which species plays the role of electron acceptor (and electron donor) extending the ordinary viewpoint of charge-transfer complexes. This can be rather simply found out by comparison of the magnitude of $\Delta\varepsilon_{\text{HO-LU}}$ obtained by the HOMO of A and the LUMO of B, and vice versa. Smaller $\Delta\varepsilon_{\text{HO-LU}}$ obtained from, say, $\varepsilon_{\text{LU(B)}} - \varepsilon_{\text{HO(A)}}$ than that from $\varepsilon_{\text{LU(A)}} - \varepsilon_{\text{HO(B)}}$ signifies that the molecule A acts as the electron donor and B as the acceptor.

2.7.1.1 Diels-Alder Reaction

Cyclization caused by addition reaction between diene and olefin in Fig. 2.66a can occur by thermal process in which the olefin is called dienophile. This kind of cycloaddition is generally called Diels-Alder reaction or [4 + 2]-cycloaddition after the numbers of π electrons involved therein. It should be stressed here that the patterns of the diene HOMO and of the dienophile LUMO are "in-phase" at the end-carbon atoms of the both molecules as demonstrated in Fig. 2.66b and vice versa as in Fig. 2.66c, the both of which would be favorable for smooth electron delocalization from the HOMO to the LUMO causing, in this case, the cycloaddition reaction in a concerted manner.

Another important factor is the smallness of the energy difference between the HOMO and the LUMO ($\Delta\varepsilon_{\text{HO-LU}}$) as mentioned in the above (ii). In most of the Diels-Alder reactions the dienes are provided with electron-donating functional group such as OMe, Me, NH_2, or NH_2Me and the dienophiles with electron-withdrawing group

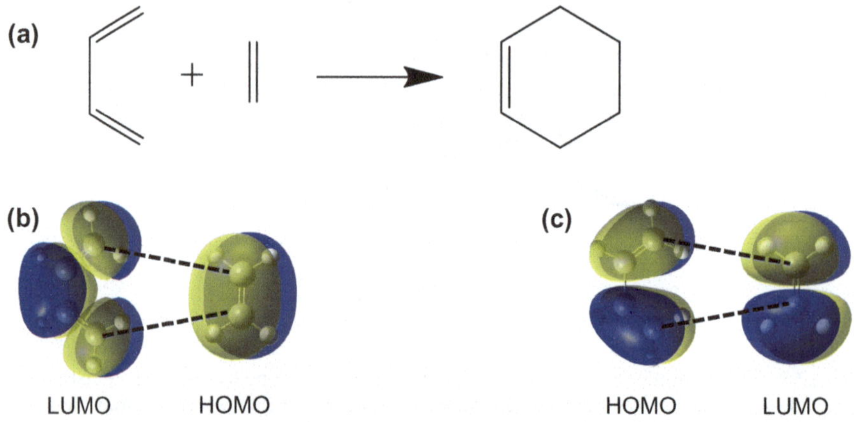

Fig. 2.66 **a** Diels-Alder reaction between *cis*-butadiene (diene) and ethylene (dienophile) and **b**, **c** HOMO-LUMO interactions between diene and dienophile based on the calculation by the HF/3-21G method. Black broken lines simply designate the in-phase orbital interaction (the same in the following drawings)

such as CHO, COOH, CN, NO_2, Ar, or halogen to accelerate the cyclization reaction. In the combination of simple diene and ethylene without functional groups, it is not yet clear which species behaves as electron donor, since the values of two kinds of $\Delta\varepsilon_{\text{HO-LU}}$ are rather close as seen in Table 2.22. Values of $\Delta\varepsilon_{\text{HO-LU}}$ for several combinations of other dienes and dienophiles substituted with typical functional group (see Fig. 2.67) are also listed in Table 2.22, where it is found in these examples that the diene should behave as electron donor and the dienophile as electron acceptor based on the smaller $\Delta\varepsilon_{\text{HO-LU}}$ value for each combination. It is also noted that small $\Delta\varepsilon_{\text{HO-LU}}$ value makes decrease in the activation energy of the reaction E_a (Sauer and Sustmann 1980).

Table 2.22 HOMO-LUMO energy difference (in au) of several combinations of dienes and dienophiles in Diels-Alder reactions based on HF/3-21G calculations

Diene	Dienophile	$\Delta\varepsilon^{a}_{\text{HO-LU}}$	Magnitude relationship	$\Delta\varepsilon^{b}_{\text{HO-LU}}$	Exp. E_a^c
cis-Butadiene	Ethylene	0.5120	≅	0.5032	27.5
cis-Butadiene	Acrolein	0.4215	$<$	0.5217	19.7
cis-Isoprene	Ethylene	0.5041	≅	0.5036	n/a
cis-Isoprene	Acrolein	0.4136	$<$	0.5221	18.7
Diazomethane[d]	Ethylene[e]	0.5040	$<$	0.5721	n/a

[a]Energy difference between the diene HOMO and the dienophile LUMO levels
[b]Energy difference between the dienophile HOMO and the diene LUMO levels
[c]Experimental value of activation energy of the reaction from Sauer and Sustmann (1980)
[d]1,3-dipole instead of diene
[e]Dipolarophile instead of dienophile

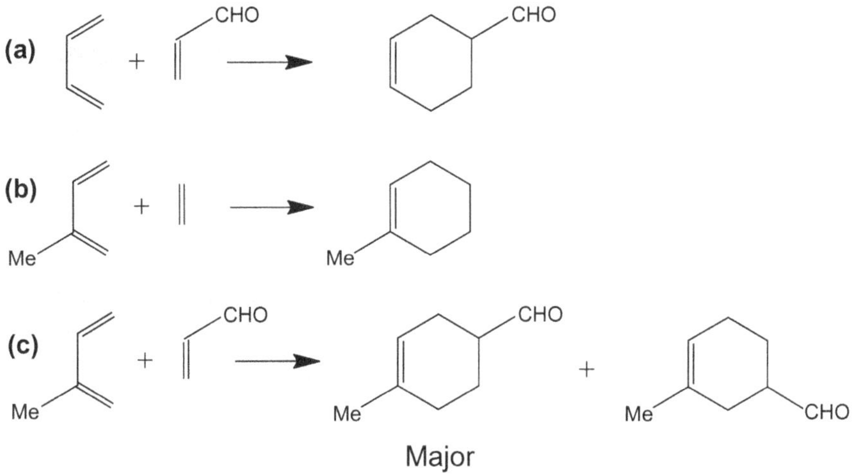

Fig. 2.67 Diels-Alder reactions between *cis*-butadiene (diene) and ethylene (dienophile) with and without functional groups. **a** *cis*-butadiene and acrolein, **b** *cis*-isoprene and ethylene, and **c** *cis*-isoprene and acrolein

It is known that reactants with functional groups at different sites result in different products as exemplified in Fig. 2.67. That is, 1-substituted diene with a substituted dienophile makes 1,2-disubstituted cyclohexene, whereas 2-substituted diene 1,4-disubstituted one. This behavior is understood by combination of the HOMO-LUMO lobe magnitudes eventually causing large stabilization of ΔE in Eq. (2.52) through similar effect to the above (iii) and is called *regioselectivity* (Houk 1973). That is, the combination of two larger MO lobes and two smaller lobes shown in Fig. 2.68 decides the specific substituted sites of the product molecule. Moreover, 1,3-dipolar addition of dipolarophile is also classified as Diels-Alder reaction. An example employing diazomethane as a 1,3-dipole and ethylene as a dipolarophile is shown in Fig. 2.69 along with the two kinds of orbital interactions mainly working therein. The $\Delta\varepsilon_{\text{HO-LU}}$ values in this reaction also listed in Table 2.22 signify that diazomethane should behave as the electron donor. Introduction of functional group to either or both of these reactants as well as change the 1,3-dipole species may change this situation.

2.7.1.2 Regioselectivity and Stereoselectivity

The first application of the orbital interaction to regioselectivity has been performed in electrophilic substitution by NO_2^+ group to naphthalene molecule as shown in Fig. 2.70a, where this reaction is experimentally known to occur at C_1 position (Fukui et al. 1952). In the electrophilic reaction, the NO_2^+ attacks the larger HOMO lobe of naphthalene as in Fig. 2.70b. Chemical reaction accompanied with, what is called, Markovnikov's rule would also be related to regioselectivity. In Fig. 2.71 is shown the reaction with a non-symmetric olefin (1-butene) and a hydrogen halide (HCl), where

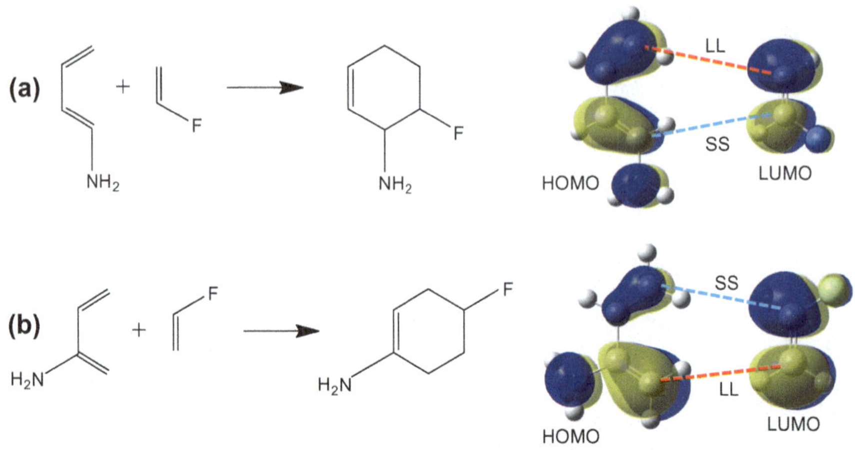

Fig. 2.68 Regioselective Diels-Alder reactions **a, b** depending on the site of functional groups. In the orbital interaction, LL designates the combination of two large MO lobes, while SS that of two small lobes throughout this section. These are based on the calculation by the HF/3-21G method

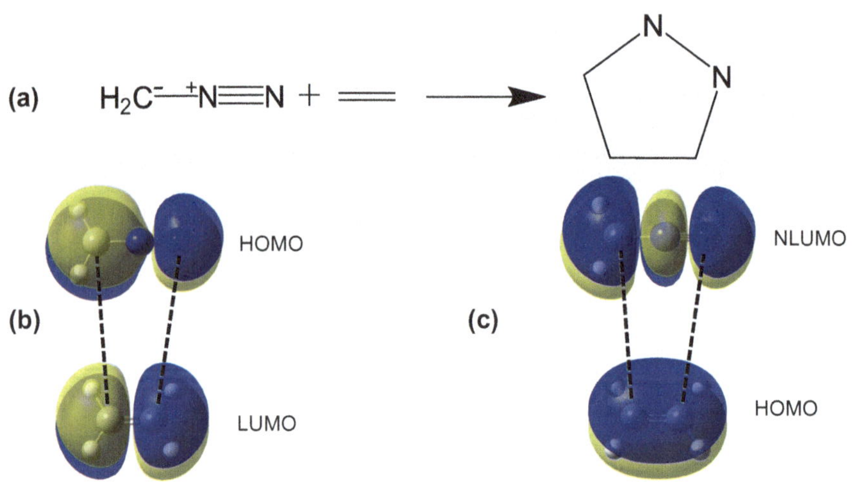

Fig. 2.69 **a** 1,3-Dipolar cycloaddition between diazomethane (1,3-dipole) and ethylene (dipolarophile) and **b, c** orbital interactions between 1,3-dipole and dipolarophile based on the calculation by the HF/3-21G method. NLUMO stands for the next LUMO considered instead of the LUMO (σ*) of diazomethane

the HOMO lobe of 1-butene is the largest at the end-carbon (C_1) and interacts with the largest LUMO lobe consisting of s-AO of proton in HCl. The electron transfer from 1-butene to HCl tends to make cleavage of H–Cl bond resulting in formation of C_2–Cl bond. Note that the reaction process including radical concerned is ruled by different regioselectivity called anti-Markovnikov's rule.

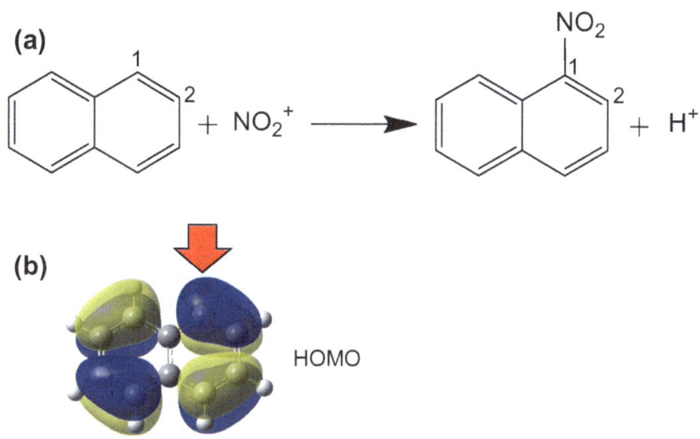

Fig. 2.70 **a** Electrophilic substitution reaction between naphthalene and NO_2^+. **b** The HOMO pattern of naphthalene showing favorable site (indicated by red arrow) in electrophilic reaction based on the calculation by the HF/3-21G method

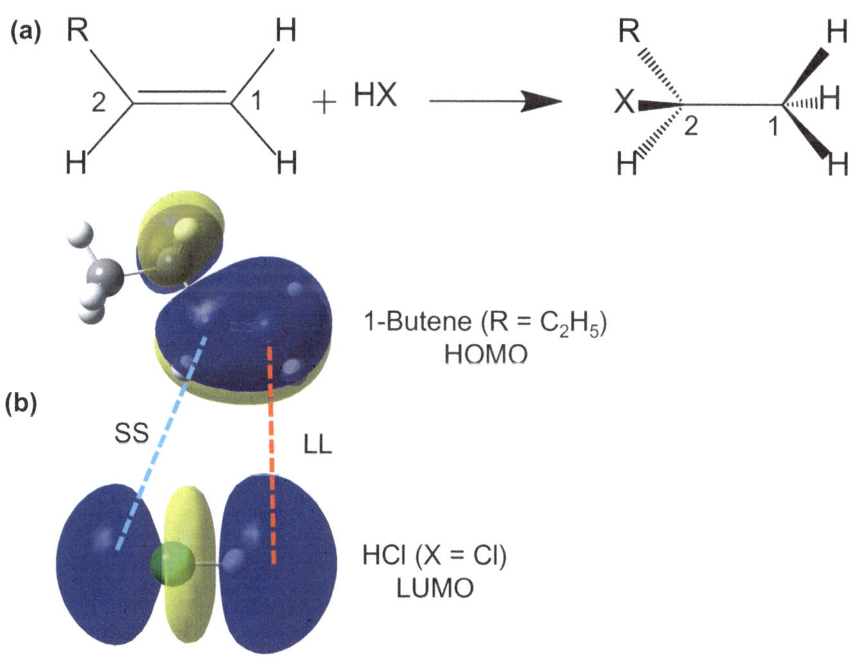

Fig. 2.71 **a** Chemical reaction between non-symmetric olefin $RCH{=}CH_2$ and hydrogen halide HX showing Markovnikov's rule and (b) HOMO-LUMO interaction between the end-carbon (C_1) of 1-butene and the proton of HCl (X = Cl) (see text) based on the calculation by the HF/3-21G method

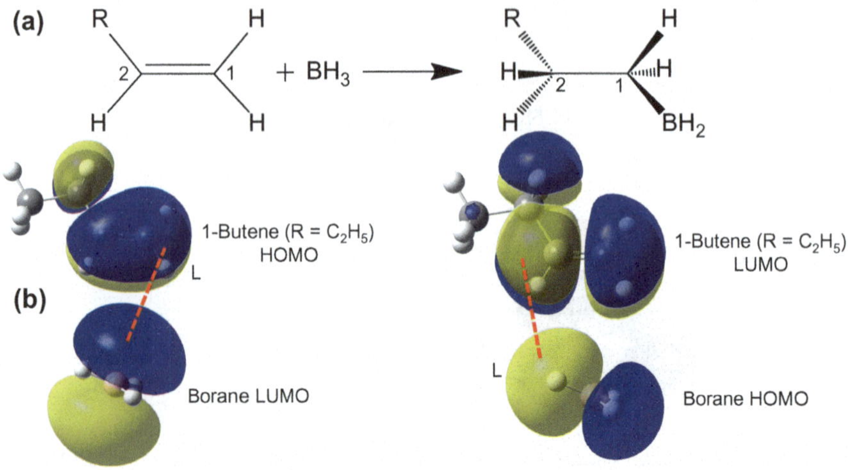

Fig. 2.72 **a** Hydroboration reaction between non-symmetric olefin and borane and **b**, **c** HOMO-LUMO interactions between 1-butene (R = C_2H_5) and borane based on the calculation by the HF/3-21G method. L signifies the large lobe

Furthermore, hydroboration (Fig. 2.72a) is another example of regioselectivity, in which the HOMO lobe of 1-butene becomes the largest at the end-carbon (C_1) and contributes to the interaction with the empty p-AO of B in borane resulting in C_1–H bond as shown in Fig. 2.72b. Successive interaction between the borane HOMO (one of the triply degenerated) lobe and the LUMO lobe of 1-butene at C_2 promotes the formation of C_2–B bond. Chemical reaction accompanied with stereoselection is also possibly interpreted by orbital-interaction examination. One of the typical examples would be classified into so-called S_N2 reaction of methyl bromide with OH^- accompanied with Walden inversion as seen in Fig. 2.73a. In this reaction, the HOMO of OH^- attacks the larger LUMO lobe stereoselectively from the backside direction of methyl bromide at the opposite side of Br as seen in Fig. 2.73b. A concomitant elimination of Br^- occurs with accompanying of the inversion of ABC atoms named by Walden inversion giving enantiomeric product as to other ABC substituents.

2.7.1.3 Radical Reactions

It is further noted that chemical reactions involving radical species give a certain variation in orbital interactions but can be similarly understood. In Fig. 2.74 are shown the MO energy diagram of a radical molecule and the ordinary closed-shell molecule, for instance, in which the singly occupied MO (SOMO)-LUMO and the HOMO-SOMO interactions possibly take place. However the SOMO-LUMO interaction is the same with the α-HOMO and the α-LUMO interaction, whereas the HOMO-SOMO one with the β-HOMO and the β-LUMO interaction in the DODS pictures as in Fig. 2.75.

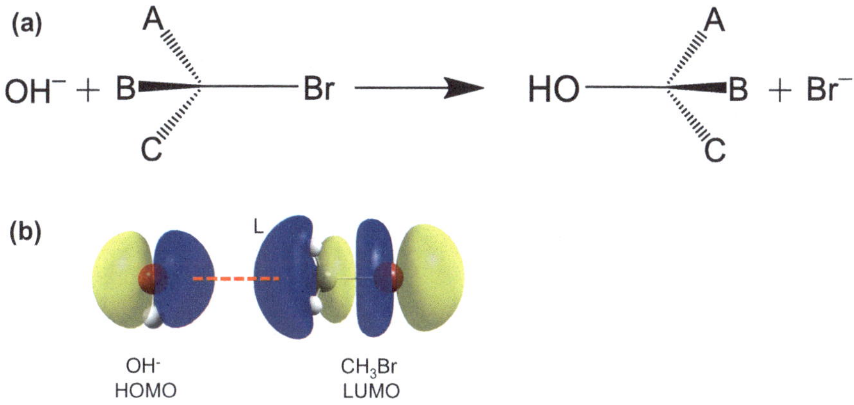

Fig. 2.73 **a** S_N2 reaction between CH_3Br (three H atoms are represented by A, B, and C to clarify the inversion) and OH^- accompanied by Walden inversion. (b) HOMO-LUMO interaction between the two species based on the calculation by the HF/3-21G method. L signifies the large lobe

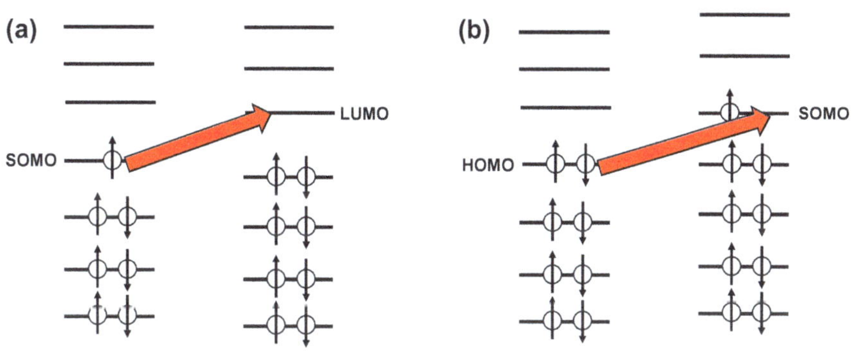

Fig. 2.74 **a** The SOMO-LUMO and **b** the HOMO-SOMO interactions

2.7.2 Reaction Path Analysis

A chemical reaction, say, from the reactant A to the product B can be theoretically dealt with on a potential energy surface (PES) mentioned in Sect. 2.1. In the reaction path analysis, PES is considered to continuously cover from the energies of the above A and B species. In theoretical chemistry, it is crucial to first find out the transition state (TS) of reacting species on the PES, activation energy, the molecular structure at the TS, and the ordinary kinetic variables accompanied. When another state such as the first-excited state participates in the reaction, a certain crossover between the concerning PES's should take place. For simplicity, we here confine ourselves to the chemical reaction in the ground state.

Definition of TS of a chemical reaction is the maximum point of the reaction path on the PES along the specific nuclear motion leading to the product in the

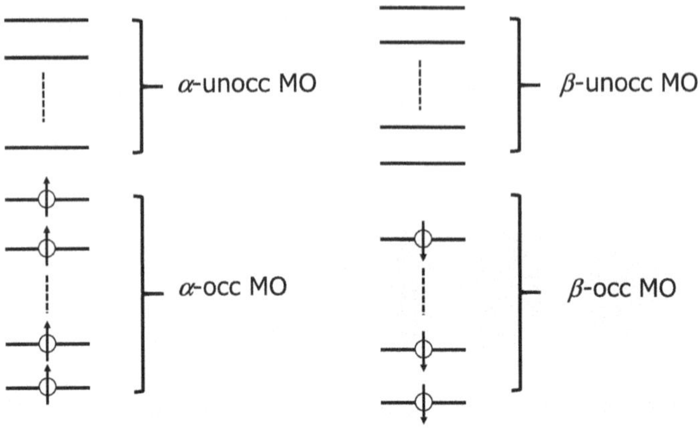

Fig. 2.75 The DODS pictures. The signs occ and unocc signify occupied and unoccupied, respectively

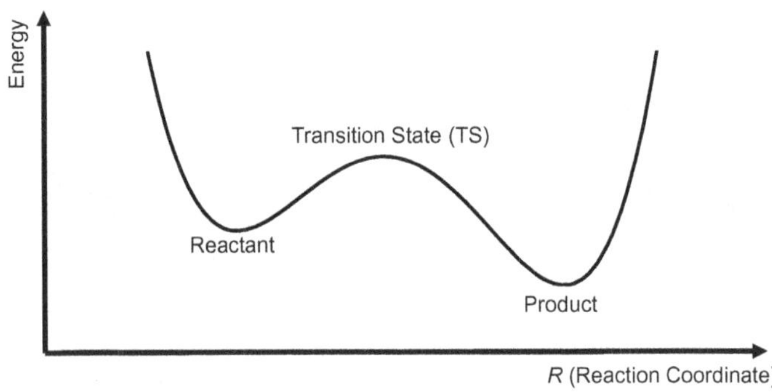

Fig. 2.76 An image of a reaction path along a reaction coordinate on the potential energy surface (PES). There are reactant, transition state, and product

direction of this particular nuclear motion direction called reaction coordinate (see Fig. 2.76). Note, however, TS corresponds to the minimum point for all other nuclear motions at the same time, which signifies that TS is defined as a saddle point or col as illustrated in Fig. 2.77. Specification of TS simultaneously affords the activation energy as well as the molecular structure of the reacting species at that state (often called activated complex), which provides information on the reaction mechanism and so forth. Process of searching for TS is also included in the problem to find a kind of stationary point on the PES but with a bit of different process from that for the optimized molecular structure mentioned in Sect. 2.1. In general, we are interested in the reaction coordinate having the TS with the smallest activation energy.

Mathematically, TS is given as the point on the PES $E(\mathbf{R})$ where all the grad E components in Eq. (2.1) become zero and that only *one* of the force constants obtained as the eigenvalues of the force constant matrix in Eq. (2.2) becomes "negative"

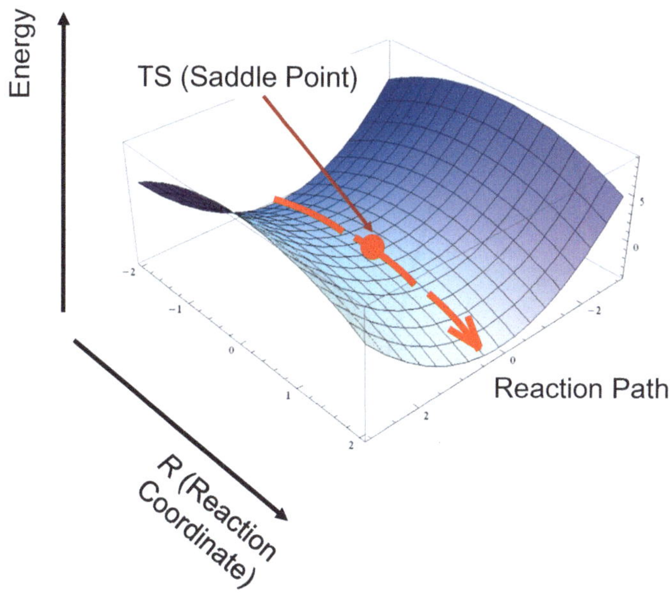

Fig. 2.77 The transition state on the reaction path (red broken line) along the reaction coordinate (R) indicated by a circle (see text)

and the others positive. The negative-force constant gives the imaginary frequency due to Eq. (2.3). The eigenvector with the negative-force eigenvalue informs us of the direction of the nuclear motion from the TS leading to the product. Actual process of finding TS had been quite manual taking rather long time and sometimes achieved even by chance. Hence it is natural that the algorithms of efficient systematic search for finding the TS should be required (Farkas and Schlegel 1999; Maeda and Ohno 2006; Schlegel 2011; Shang and Liu 2012). For instance, a methodology of finding of the TS starting from an arbitrary equilibrium structure represented by a local minimum on the PES during a chemical reaction has been developed (Ohno and Maeda 2004). Actual examples of the TS's found by this method during an isomerization reaction from hexasilabenzene molecule to hexasilaprismane are illustrated in Fig. 2.78 (Moteki et al. 2009). There are seen four TS's during this reaction, which should provide interesting information for the chemical reaction analyses.

The intrinsic reaction coordinate (IRC) can be defined for the reaction path with the steepest descent from the TS to the product with infinitesimally small velocity (Fukui 1970, 1981). Theoretically, there can be two IRC's from the TS to both the reactant and the product, where it is normally considered that both these two correspond to the steepest ascent and descent paths. The IRC often affords prediction of the mechanism of chemical reaction. For instance, in Fig. 2.79, the result of the IRC analysis for the isomerization process of diademane to triquinacene is shown. Both the negative magnetic susceptibility coming from diamagnetic property and

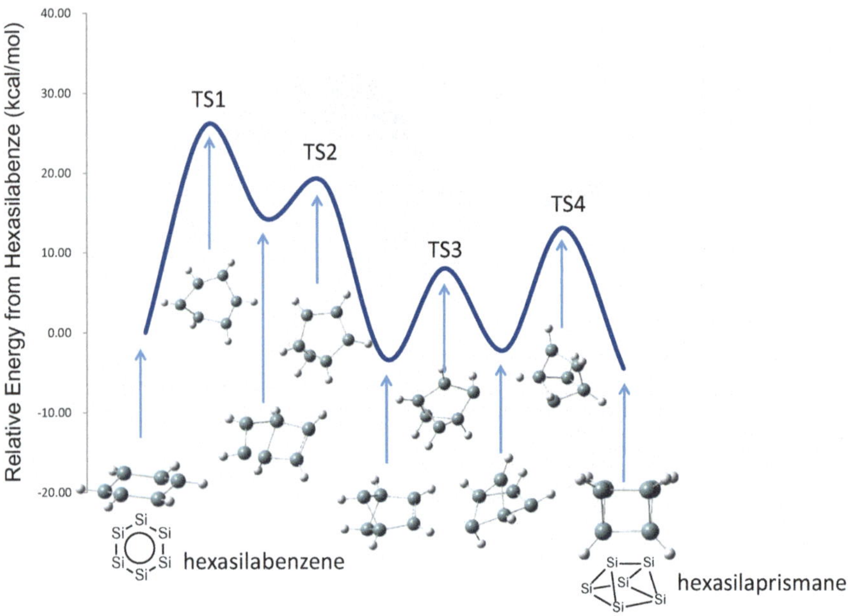

Fig. 2.78 Transition states (TS's) during the isomerization reaction from hexasilabenzene to hexasilaprismane obtained by DFT/B3LYP/6-311G**. Adapted with permission from Moteki et al. (2009). Copyright 2009 American Chemical Society

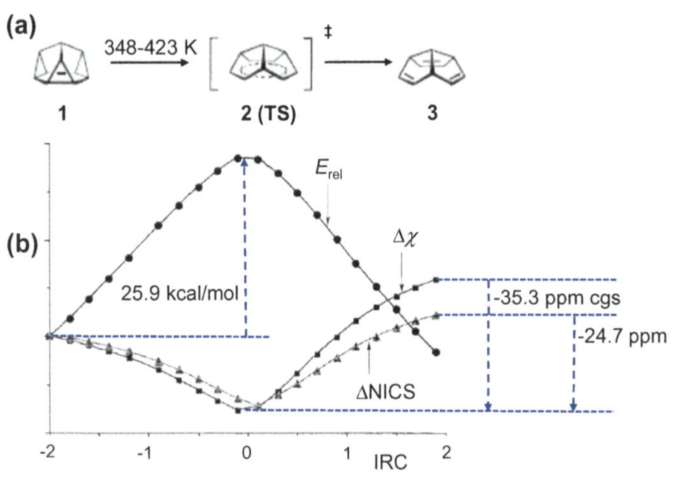

Fig. 2.79 IRC search results of **a** isomerization process of diademane 1 into triquinacene 3 where 2 signifies the transition state. **b** Relative energy, change in magnetic susceptibility, and NICS value along the IRC −2, 0, 2 correspond to **1**, **2** (TS), and **3**, respectively) based on the calculation by the HF/6-31G* (for relative energy) and HF GSGT/6-31G*//HF/6-31G* (for magnetic susceptibility and NICS) methods. Reprinted with permission from Verevkin et al. (1998). Copyright 1998 American Chemical Society

the negative NICS (nuclear-independent chemical shift: see Sect. 2.8.2) value imply the aromatic characteristics of the TS molecule. There have also been developed a couple of automated algorithms for the IRC search such as, for instance, the "anharmonic downward distortion following" (ADDF) (Ohno and Maeda 2004) and the "artificial force induced reaction" (AFIR) (Maeda et al. 2015) methods. Once such the reaction path on the PES connecting the initial local minimum (reactant), TS, and the final local minimum (product) is established, the reaction rate constant for the corresponding elementary reaction can be evaluated by the transition-state theory (Fernández-Ramos et al. 2006).

2.8 Molecular Design

This section actually shows a couple of examples of molecular design toward extracting intriguing electronic properties from molecules and polymers. Such material is possibly useful to fabricate molecular devices. The molecular design introduced here utilizes controlling of spin alignment, band-gap value, and fluorescence efficiency.

2.8.1 High-Spin Molecules

Most of the electronic structures of organic molecules favor the lowest possible spin multiplicity as the ground state due to the Pauli principle. For instance, molecules consisting of even numbers of electrons usually become energetically most stable (ground state) at the singlet spin state, and of odd numbers at the doublet spin state. Hence it is rather unusual for organic molecules to have high-spin state (see, e.g., Fig. 2.80) as the ground state. In other words, ferromagnetic property hardly appears in organic molecules or polymers except for those containing inorganic elements with d or f electrons acting as the spin centers. In this sense, theoretical design of small organic molecules with a stable high-spin state, such as triplet state or higher ones, is intriguing and rather challenging as the starting point to explore organic oligomers and/or polymers with higher spin state than the singlet state.

It would be useful to point out here that there are simple rules to predict the stable high-spin state for non-Kekulé-type molecules that cannot be expressed by classical Kekulé structures. In alternant hydrocarons let us label the conjugated atoms as starred and unstarred ones alternatively as shown in Fig. 2.81 so that no two atoms with the same labels are directly linked. Putting the numbers of starred and unstarred carbon atoms in an alternant hydrocarbon molecule as n^* and n, respectively, it can be predicted that this molecule has $(n^* - n)$ non-bonding MO's (NBMO's) and $(n^* - n)$ unpaired electrons as well (Dewar 1969). The total spin number S of the molecules in Fig. 2.81 is thus expressed by

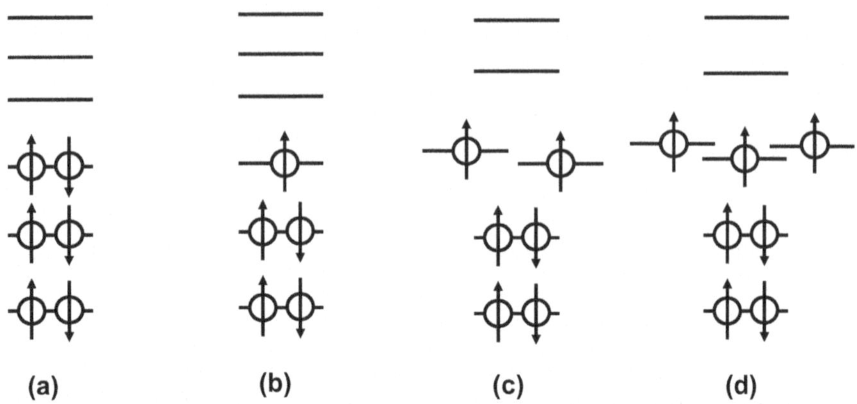

Fig. 2.80 Schematic representations of MO energy levels of **a** closed-shell singlet, **b** open-shell doublet, **c** nearly triplet, and **d** nearly quartet states. The arrows with two directions indicate α and β spins as usual. Note the SOMO's in **c** and **d** are almost energetically degenerate or pseudo-degenerate

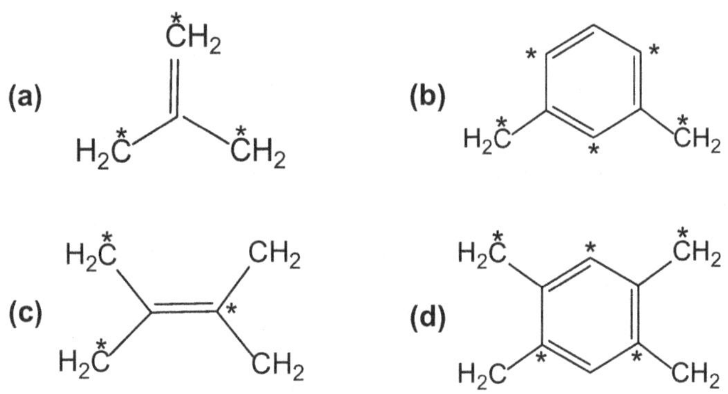

Fig. 2.81 **a** Trimethylenemethane (TMM), **b** *m*-quinodimethane (MQDM), **c** tetramethyleneethylene (TME), and **d** 1,2,4,5-tetramethylenebenzene (TMB) with starred and unstarred carbon atoms

$$S = \frac{n^* - n}{2} \tag{2.53}$$

(Ovchinnikov 1978) as listed in Table 2.23. It is seen, hence, trimethylenemethane (TMM) and *m*-quinodimethane (MQDM) would have the triplet-ground state but tetramethyleneethylene (TME) and 1,2,4,5-tetramethylenebenzene (TMB) would not. This simple rule will basically work in what follows and, furthermore, in Sect. 4.1.3.

More sophisticated ideas toward design of high-spin molecules come from utilization of molecular symmetry causing degeneracy of the SOMO's each of which accommodates an electron spin. The degeneracy or pseudo-degeneracy of the SOMO

Table 2.23 Count of spin quantum numbers for several non-Kekulé molecules with numbers of starred and unstarred carbon atoms ($n^*$and n)

	n^*	n	Spin quantum number S (=[$n^* - n$]/2)	Multiplicity: $2S$+1
TMM	3	1	1	3; triplet
MQDM	5	3	1	3; triplet
TME	3	3	0	1; singlet
TMB	5	5	0	1; singlet

levels guarantees Hund's rule resulting in parallel spin alignment (Ito et al. 1997). For instance, a benzene ring equipped with two radical centers at its meta position could have triplet-ground state or, in other words, the triplet state becomes energetically most stable as expressed by positive ΔE_{ST} value defined by

$$\Delta E_{\mathrm{ST}} = E_{\mathrm{S}} - E_{\mathrm{T}} \tag{2.54}$$

where E_{S} and E_{T} represent the energies of a molecule in the singlet and triplet states, respectively. Two kinds of molecular designs satisfying these conditions are exemplified in Figs. 2.82 and 2.83 (MQDM and *m*-phenylenediamine dication (MPD^{2+})). Note that MPD^{2+} is an isoelectronic molecule of MQDM. It is seen that these two

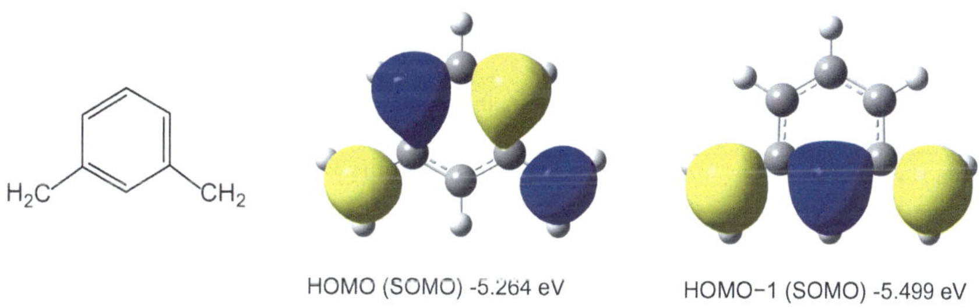

Fig. 2.82 Two SOMO patterns of MQDM with orbital energies. Note that these SOMO's are nearly degenerated

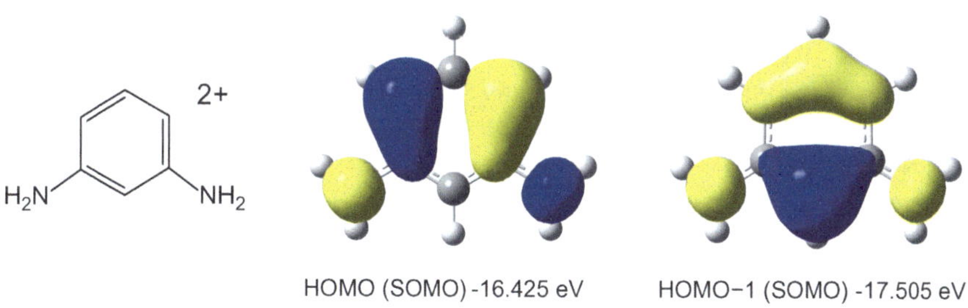

Fig. 2.83 Two SOMO patterns of MPD^{2+} with orbital energies. Note that these SOMO's are nearly degenerated

molecules have two SOMO's being nearly degenerated. As a matter of fact, the calculated ΔE_{ST} values of these become positive guaranteeing stability of the triplet state as listed in Table 2.24. Atomic spin densities listed in Table 2.25 show alternative spin alignment due to spin polarization in the both MQDM and MPD^{2+} where C^1, C^3, C^5, C^{11} (N^{11} in MPD^{2+}), and C^{12} (N^{12} in MPD^{2+}) possess large α spins as expected in this molecular design. It is rather interesting that N^{11} and N^{12} in MPD^{2+} have negative atomic net charges in spite of dicationic species. Instead, the hydrogen atoms H^7, H^9, and H^{10} attached to nitrogen atoms are rather protonic, the tendency of which is also seen in MQDM. Distribution of the atomic spin densities is shown in Fig. 2.84. Basic idea presented here can be applicable to molecular design toward higher spin state. In fact, attempt to obtain higher spin state such as quartet state has been published elsewhere (Yoshizawa et al. 1993; Ito et al. 1998). Further extensive

Table 2.24 Comparison of the singlet and triplet states of MQDM and MPD^{2+} based on DFT/B3LYP/6-31+G** calculations

	Multiplicity	Symmetry	Electronic state	ΔE_{ST} (kcal/mol)[a]
MQDM	Closed-shell singlet	C_{2v}	$^1A_1$	29.50
	Open-shell triplet	C_{2v}	$^3B_2$	
MPD^{2+}	Closed-shell singlet	C_{2v}	$^1A_1$	8.53
	Open-shell triplet	C_{2v}	$^3B_2$	

[a]Positive value signifies stabilization of the triplet state (see text)

Table 2.25 Atomic net charges and atomic spin densities of MQDM and MPD^{2+} based on DFT/B3LYP/6-31+G** calculations[a]

Atom No.	MQDM		MPD^{2+}	
	Atomic net charge	Atomic spin density	Atomic net charge	Atomic spin density
C^1, C^3	−0.2018	0.3801	−0.1226	0.3863
C^2	−0.2585	−0.1637	−0.1672	−0.1070
C^4, C^6	−0.1449	−0.1780	0.1888	−0.0340
C^5	−0.1642	0.4491	−0.0388	0.6171
H^7, H^9	0.2415	−0.0116	0.3112	−0.0116
H^8	0.2497	0.0047	0.3289	0.0027
H^{10}	0.2351	−0.0135	0.2940	−0.0178
C^{11}, C^{12} (N^{11}, N^{12})[b]	−0.3604	0.7103	−0.5685	0.4344
H^{13}, H^{16}	0.2173	−0.0196	0.4944	−0.0114
H^{14}, H^{15}	0.2171	−0.0195	0.4881	−0.0112
Total	0	2	2	2

[a]From natural population analyses
[b]For MPD^{2+}

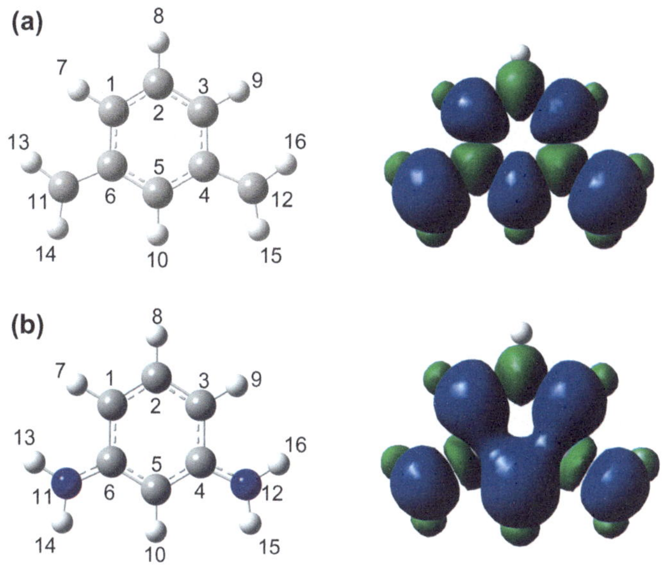

Fig. 2.84 Atomic numberings and spin densities of **a** MQDM and **b** MPD^{2+}. In the spin densities, blue colors indicate α-spins and green β-spins

description along this theme is to be given in Sect. 4.1.3 where this idea is more developed toward nanomaterials possessing higher spin state such as septet or more.

2.8.2 *Narrow Band-Gap Polymers*

It is well known that the electrically conductive polymer or organic conductive crystal should have zero bandgap (see Sect. 3.3 as to details of bandgap). However, apart from conductive polymers showing metallic property, those with narrow bandgap are also useful and of interest due to their application to molecular devices such as semiconductors toward organic FET (field-effect transistor), LED (light-emitting diode), and so on.

For instance, polyacene (PAc), with an infinite 1D array of condensed aromatic ring, shown in Fig. 2.85a is a member of, what is called, graphene nanoribbons (GNR's). Oligoacenes such as pentacene is a member of PAc with finite length. Electronic structure of PAc can be dealt with 1D CO method, which will be described in Sect. 3.3, and gives the corresponding band structure as shown in Sect. 3.3.4 for polythiopene (PT) case. PAc has rather narrow bandgap (0.504 eV) based on the early CO calculation at the semiempirical HF calculation level (Tanaka et al. 1983).

On the contrary, PPh being a structural isomer of PAc (see Fig. 2.85b) has rather larger band-gap value of 10.636 eV (Tanaka et al. 1983). This difference in the bandgaps is considered to come from the degree of aromaticity being the larger in

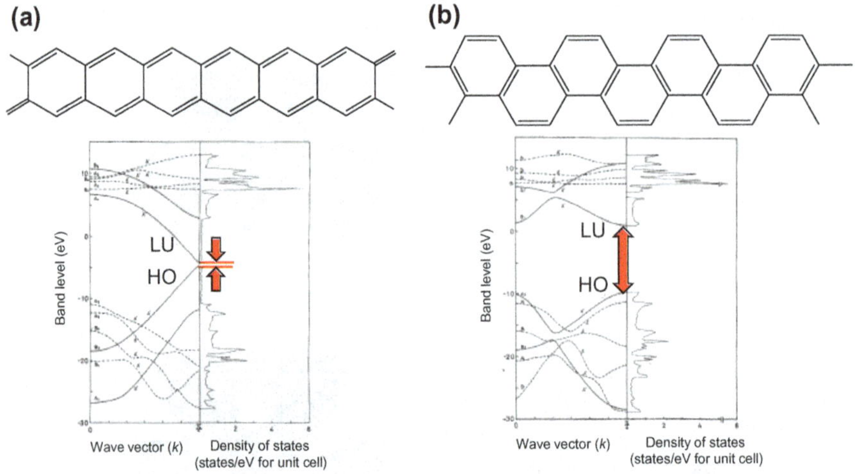

Fig. 2.85 Molecular structures and band structures of **a** polyacene (PAc) and **b** polyphenanthrene (PPh) based on /crystal orbital/semiempirical (CNDO/2) calculations. HO and LU stand for the highest occupied and the lowest unoccupied bands, respectively. The red arrows signify the bandgap. Reprinted from Tanaka et al. (1983), Copyright 1983, with permission from Elsevier

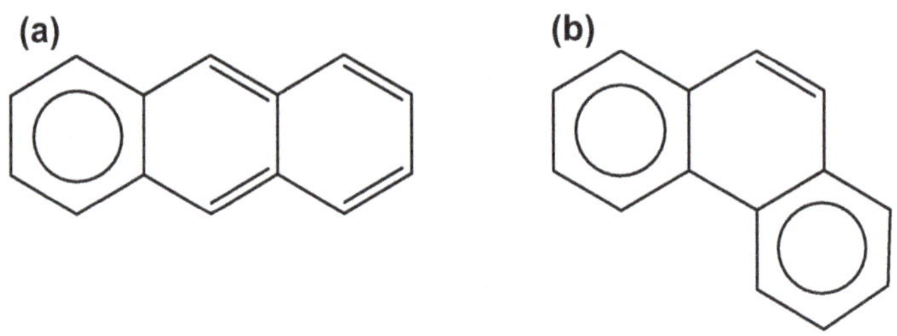

Fig. 2.86 Numbers of writable sextet structures in **a** anthracene and **b** phenanthrene

PPh. A simple explanation as to such difference would be possible by comparison of numbers of writable sextet structures in each molecule (Clar 1972). That is, as illustrated in Fig. 2.86, there can be written one sextet structure in anthracene while two in phenanthrene. Further interesting discussion on this property using the concept of the "schnitt" of MO's has been given (Fukui 1982). As a matter of fact, concept of aromaticity needs profound consideration, since there is no physical substance of that and only indirect phenomena such as diamagnetic susceptibility and NMR chemical shift. Incidentally, the nucleus-independent chemical shift (NICS) is often used instead of the ordinary NMR chemical shift in recent years for the sake of convenience (Schleyer et al. 1996), where a negative NICS value is considered to represent aromatic characteristics.

Furthermore, PT mentioned above is one of the most popular electrically conductive polymers when it is doped with p-type dopant to introduce positive holes to the polymer main chain. The intrinsic PT before doping has a band-gap value ΔE_g of 1.965 eV as listed in Table 2.26. It will be possible to obtain polymers of narrower bandgaps by considering polythiophene oxide (PTO) and polythiophene dioxide (PTDO) by changing S atom in the five-membered ring of PT into sulfoxide (SO) and sulfonyl (SO_2) groups, respectively, as shown in Fig. 2.87 (Tanaka et al. 1989). That is, introduction of SO or SO_2 group has been supposed to result in a new HO level between the original HO and LU levels of PT due to molecular design based on the orbital-interaction analyses. Explanation of this situation with respect to PTO would be done as illustrated in Fig. 2.88.

It is suggested that formation of the HOMO-1 and the LUMO of thiophene molecule comes from the orbital interaction between the π-HOMO of *cis*-butadiene frame and 3pπ AO of S atom considered by virtual intramolecular separation of a thiophene moiety as shown in Fig. 2.88a. The HOMO of thiophene molecule, on the other hand, comes from the LUMO of the *cis*-butadiene in an intact manner. The frontier MO's of thiophene molecule are thus made up and the CO's of PT in a similar manner. This situation becomes different by introduction of SO group instead of an S atom as shown in Fig. 2.88b. That is, antibonding orbital interaction between the HOMO of the *cis*-butadiene and the π MO of the SO group makes the new HOMO with the energy level between those of the original HOMO and the LUMO of the thiophene molecule. This also changes the original HOMO of the thiophene molecule into the new LUMO of thiophene oxide and the CO's of PTO in a similar manner.

Band structures of these polymers in Fig. 2.89 based on the CO calculations by DFT/B3LYP/6-31G** give the ΔE_g values as listed in Table 2.26, in which it is seen that those of PTO and PTDO actually becomes narrower than that of PT.

Table 2.26 Band-gap (ΔE_g) values of PT, PTO, and PTDO

	PT	PTO	PTDO
ΔE_g (eV)	1.965	0.931	0.964

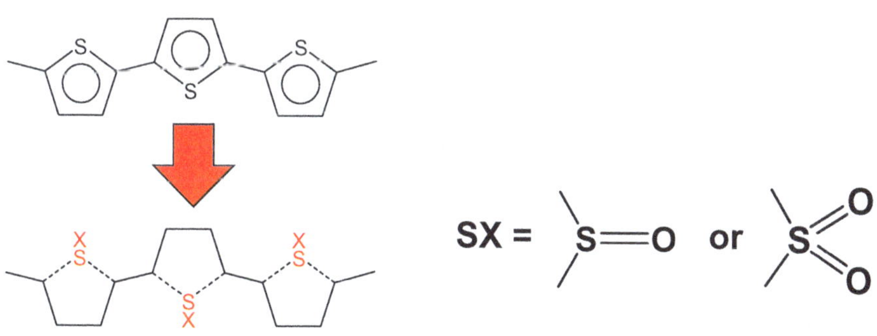

Fig. 2.87 Modification of polythiophene (PT) by replacing S atom with SO (sulfoxide) or SO_2 (sulfonyl) group

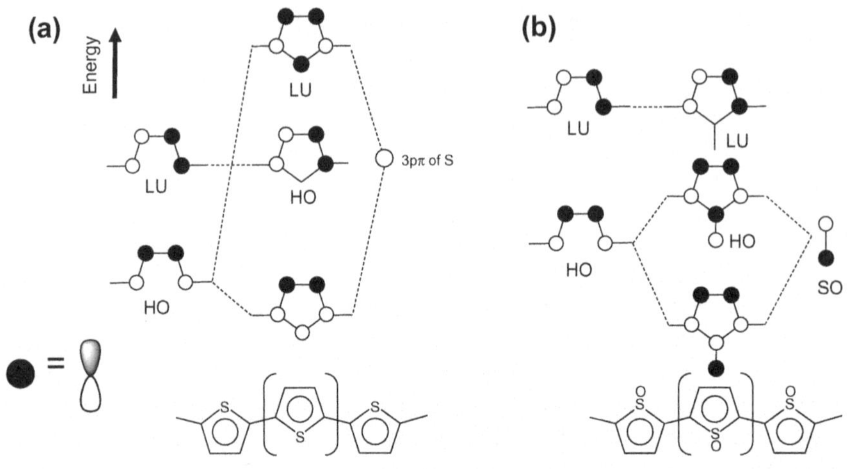

Fig. 2.88 Orbital-interaction analysis of lowering the LU level of PT by replacement of S with SO group. Reprinted from Tanaka et al. (1983), Copyright 1983, with permission from Elsevier

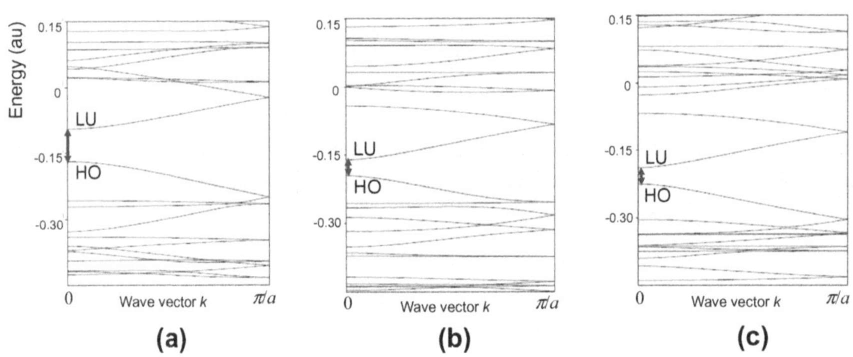

Fig. 2.89 Band structures of **a** PT, **b** PTO, and **c** PTDO based on the DFT/crystal orbital/B3LYP/6-31G**. The red arrows signify the bandgap

2.8.3 *Thermally Activated Delayed-Fluorescence (TADF) Molecule*

Fluorescent molecules working by injection of external electron and hole is a key material as photo-generation layer in organic light-emitting diode (OLED) device shown in Fig. 2.90, and molecular design of such molecules with high efficiency is being paid attention to in recent years. The injected holes and electrons from the both electrodes eventually construct singlet exciton (or electron-hole) pairs like in the singlet-excited state (usually S_1) to emit fluorescence from the photo-generation layer. It is understood, however, that 75% of the S_1 population would turn into the triplet state (T_1) through the intersystem crossing (ISC) (Helflich and Schneider

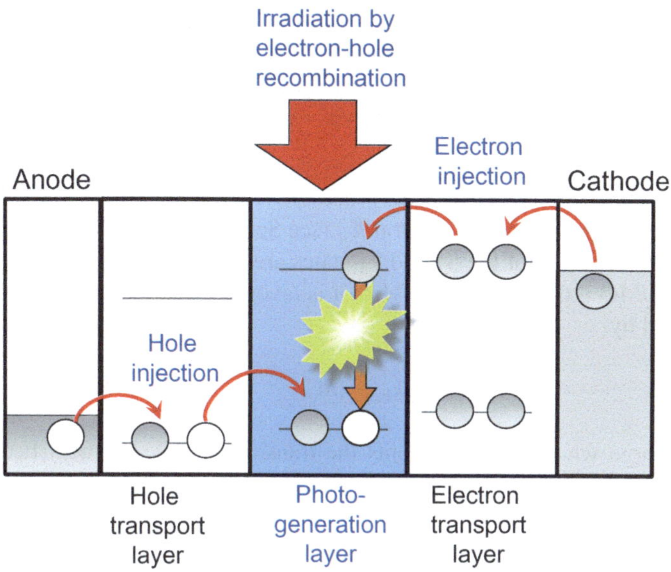

Fig. 2.90 Schematic diagram of organic light-emitting diode (OLED) device, in which all the layers consist of organic molecules

1965). Usual organic molecules do not give phosphorescence from the T_1 state but nonradiative transition. Hence efficient utilization of the T_1 state is desirable for higher efficient photo radiation.

One of such strategies would be to employ thermally activated delayed-fluorescence (TADF) process as illustrated in Fig. 2.91. That is, thermal activation from the T_1 back to the S_1 state is intended and designed, so as to re-enhance the number of electrons into the S_1 state using the reverse ISC (RISC) process from the T_1 to the S_1 state. The new S_1 state will afford delayed fluorescence compared with the original one, which eventually increases the fluorescence efficiency. In order to

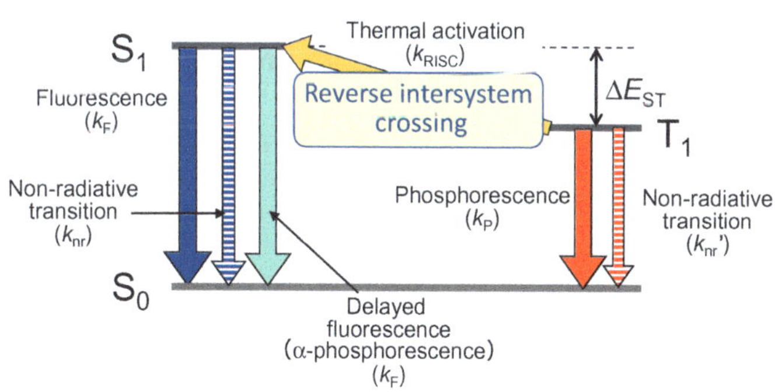

Fig. 2.91 Diagram of thermal activated delayed fluorescence (TADF)

enhance the TADF process, the energy difference between T_1 and S_1 states defined by ΔE_{ST} ought to be small due to the relationship (Berberan-Santos and Garcia 1996)

$$k_{RISC} \propto \exp[-\Delta E_{ST}] \tag{2.55}$$

where k_{RISC} stands for the rate of RISC process as shown in Fig. 2.91. The value ΔE_{ST} is determined by the exchange integral K_{ij} (see Sect. 3.1.1) under the condition that electron configurations of the S_1 and T_1 states are mostly decided by the transition from the HOMO i to the LUMO j. Also, it is favorable that the rate of fluorescence k_F expressed by

$$k_F \propto |\mathbf{m}_{01}|^2 \tag{2.56}$$

should be large where $\mathbf{m}_{01}$ represents the transition dipole moment between the S_0 and the S_1 states to suppress the nonradiative transition (Toptygin 2003). This moment can be simplified into $\mathbf{m}_{ij}$ between the HOMO i and the LUMO j again under the condition that electron configuration of the S_1 state is mostly decided by the transition from the HOMO i to the LUMO j.

These requirements have been managed to realize by considering molecular design, e.g., using N^1-[4-(4,6-diphenyl-1,3,5-triazin-2-yl)phenyl]-N^1-[4-(diphenylamino)phenyl]-N^4,N^4-diphenylbenzene-1,4-diamine (DPA-TRZ) as shown in Fig. 2.92 (Shizu et al. 2015a). That is, it is devised, in this molecule, plane of the TRZ motif in the right-hand side (indicated by blue) is considerably

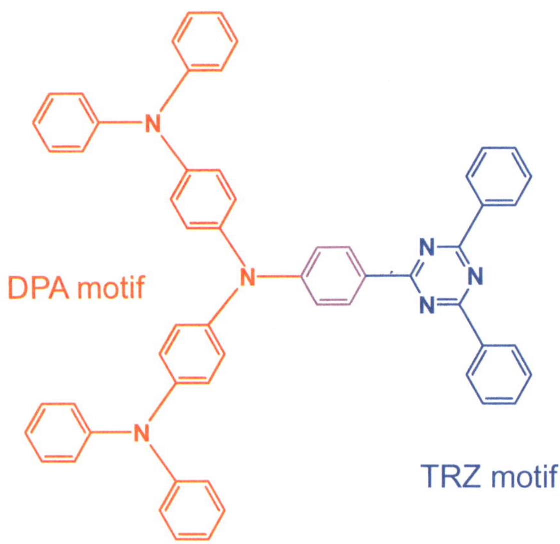

Fig. 2.92 N^1-[4-(4,6-diphenyl-1,3,5-triazin-2-yl)phenyl]-N^1-[4-(diphenylamino)phenyl]-N^4,N^4-diphenylbenzene-1,4-diamine (DPA-TRZ) molecule based on the molecular design. Reprinted with permission from Shizu et al. (2015a). Copyright 2015 by the American Physical Society

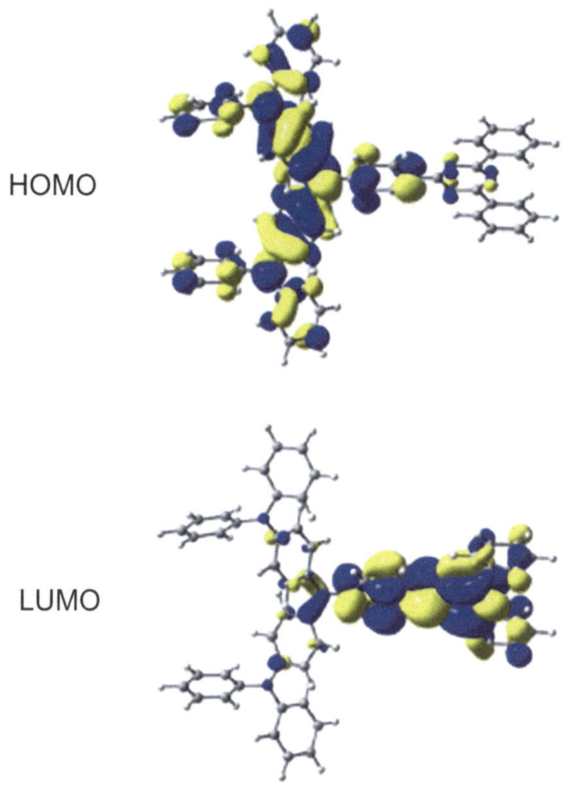

Fig. 2.93 The HOMO and the LUMO patterns of DPA-TRZ molecule based on the TD-M06-2X-cc-pVDZ calculation. Reprinted with permission from Shizu et al. (2015a). Copyright 2015 by the American Physical Society

twisted from that of the DPA in the left (indicated by red) to suppress the ΔE_{ST} value while remaining the value of $|\mathbf{m}_{01}|$ as much as possible by inspection of the detailed contents of the expressions of ΔE_{ST} and $|\mathbf{m}_{01}|$. This DPA-TRZ molecule has been designed so that intramolecular charge transfer may easily take place from the DPA to the TRZ motif. This is understood by that the HOMO is localized at the left side and the LUMO at the right side as shown in Fig. 2.93. Note that these two motifs are rather twisted in order to obtain a small ΔE_{ST} value (0.11 eV). The external quantum efficiency of fabricated OLED using this molecule reaches 13.8%. Furthermore, another improved molecule, 9-(4-(4,6-diphenyl-1,3,5-triazin-2-yl)phenyl)-9′-phenyl-9*H*,9′*H*-3,3′-bicarbazole (BCzT) shown in Fig. 2.94 has achieved the external quantum efficiency of 21.7% (Shizu et al. 2015b).

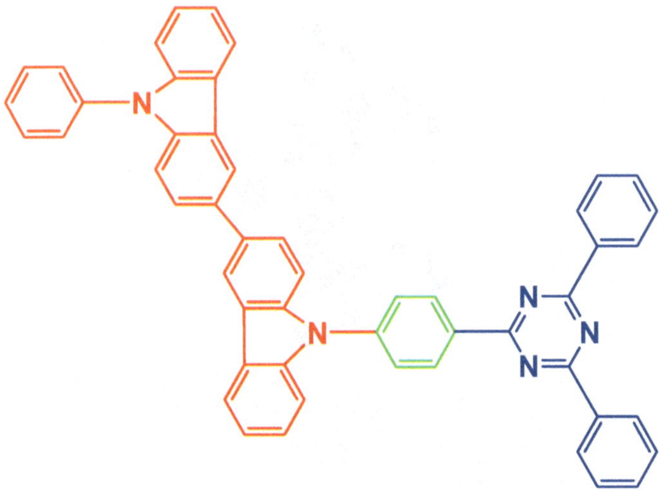

Fig. 2.94 9-(4-(4,6-diphenyl-1,3,5-triazin-2-yl)phenyl)-9′-phenyl-9*H*,9′*H*-3,3′-bicarbazole (BCzT) molecule based on the molecular design. Adapted with permission from Shizu et al. (2015b). Copyright 2015 American Chemical Society

References

A. Ambrosetti, P. L. Silverstrelli, J. Phys. Chem. C **115**, 3695–3702 (2011)

R.F.W. Bader, Chem. Rev. **91**, 893–928 (1991)

F. Bartha, F. Bogár, A. Peeters, C. van Alsenoy, V. van Doren, Phys. Rev. B **62**(10), 142–150 (2000)

M.N. Berberan-Santos, J.M.M. Garcia, J. Am. Chem. Soc. **118**, 9391–9394 (1996)

A.D. Buckingham, Quart. Rev. Chem. Soc. **13**, 183–214 (1959)

A.D. Buckingham, B.J. Orr, Quart. Rev. Chem. Soc. **21**, 195–212 (1967)

C. van Caillie, R.D. Amos, Chem. Phys. Lett. **308**, 249–255 (1999)

M. Carnell, S. Grimme, S.D. Peyerimhoff, Chem. Phys. **179**, 385–394 (1994)

J. Cioslowski, *Electronic Structure Calculations on Fullerenes and Their Derivatives* (Oxford University Press, Oxford, 1995)

E. Clar, *The Aromatic Sextet*, Chap. 3 (Wiley, New York, 1972)

J.B. Collins, A. Streitwieser Jr., J. Comput. Chem. **1**, 81–87 (1980)

J.D. Cox, G. Pilcher, *Thermochemistry of Organic and Organometallic Compounds* (Academic Press, New York, 1970)

S. Dasgupta, W.B. Hammond, W.A. Goddard III, J. Am. Chem. Soc. **118**, 12291–12301 (1996)

W.I.F. David, R.M. Ibberson, J.C. Mathewman, K. Prassides, T.J.S. Dennis, J.P. Hare, H.W. Kroto, R. Taylor, D.R.M. Walton, Nature, **353**, 147–149 (1991).

M.J.S. Dewar, *The Molecular Orbital Theory of Organic Chemistry*, Chap. 6 (McGraw-Hill, New York, 1969)

C. Edmiston, K. Ruedenberg, Rev. Mod. Phys. **35**, 457–465 (1963)

Ö. Farkas, H.B. Schlegel, J. Chem. Phys. **111**, 10806–10814 (1999)

A. Fernández-Ramos, J.A. Miller, S.J. Klippenstein, D.G. Truhlar, Chem. Rev. **106**, 4518–4584 (2006)

W.H. Flygare, R.C. Benson, Mol. Phys. **20**, 225–250 (1971)

J.B. Foresman, M. Head-Gordon, J.A. Pople, M.J. Frisch, J. Phys. Chem. A **96**, 135–149 (1992)

J.M. Foster, S.F. Boys, Rev. Mod. Phys. **32**, 300–302 (1960)

K. Fukui, T. Yonezawa, H. Shingu, J. Chem. Phys. **20**, 722–725 (1952)
K. Fukui, J. Phys. Chem. **74**, 4161–4163 (1970)
K. Fukui, Acc. Chem. Res. **4**, 57–64 (1971)
K. Fukui, Acc. Chem. Res. **14**, 363–368 (1981)
K. Fukui, *LES PRIX NOBEL EN 1981* (Nobel Lecture in 1981), (1982), pp. 146–166
P. George, M. Trachtman, C.W. Bock, A.M. Brett, Tetrahedron **32**, 317–323 (1976)
S.J. Grabowski, Chem. Rev. **111**, 2597–2625 (2011)
M. Gussoni, R. Rui, G. Zerbi, J. Mol. Struct. **447**, 163–215 (1998)
J.C.L. Hageman, R.J. Meier, M. Heinemann, R.A. de Groot, Macromolecules **30**, 5953–5957 (1997)
A.E. Hansen, T.D. Bauman, Adv. Chem. Phys. **44**, 545–644 (1980)
W. Helfrich, W.G. Schneider, Phys. Rev. Lett. **14**, 229–231 (1965)
T. Helgaker, P. Jørgensen, J. Chem. Phys. **95**, 2595–2601 (1991)
K.-H. Hellwege, A.M. Hellwege (eds.), *Landolt-Börnstein, Numerical Data and Functional Relationships in Science and Technology, New Series, II/6* (Springer, Berlin, 1974)
K.-H. Hellwege, A.M. Hellwege (eds.), *Landolt-Börnstein, Numerical Data and Functional Relationships in Science and Technology, New Series, II/14a* (Springer, Berlin, 1982)
G. Henkelman, A. Arnaldsson, H. Jónsson, Comput. Mat. Sci. **36**, 354–360 (2006)
R. Hoffmann, J. Chem. Phys. **39**, 1397–1412 (1963)
K.N. Houk, J. Am. Chem. Soc. **95**, 4092–4094 (1973)
J.B. Huntington, *Solid State Physics*, vol. 7, ed. by F. Seitz, D. Turnbull (Academic, New York, 1958)
T. Ikeda, H. Adachi, H. Fueno, K. Tanaka, T. Haino, J. Org. Chem. **82**, 10062–10069 (2017)
A. Ito, H. Miyajima, K. Yoshizawa, K. Tanaka, T. Yamabe, J. Org. Chem. **62**, 38–43 (1997)
A. Ito, A. Taniguchi, K. Yoshizawa, K. Tanaka, T. Yamabe, Bull. Chem. Soc. Jpn **71**, 337–343 (1998)
L. Itzhaki, E. Altus, H. Basch, S. Hoz, Angew. Chem. Int. Ed. **44**, 7432–7435 (2005)
T. Iwamoto, M. Tamura, C. Kabuto, M. Kira, Science **290**, 504–506 (2000)
J.F. Janak, Phys. Rev. B **103**, 7165–7168 (1978)
K. Kitaura, K. Morokuma, Intern. J. Quantum Chem. **10**, 325–340 (1976)
M. Kobayashi, N. Hayakawa, T. Matsuo, B. Li, T. Fukunaga, D. Hashizume, H. Fueno, K. Tanaka, K. Tamao, J. Am. Chem. Soc. **138**, 758–761 (2016)
T. Koopmans, Physica **1**, 104–113 (1934)
P. Krishnakumar, D.K. Maity, Comp. Theor. Chem. **1099**, 185–194 (2017)
H.W. Kroto, J.R. Heath, S.C. O'Briene, R.F. Curl, R.E. Smalley, Nature **318**, 162–163 (1985)
K. Larsson, Ark. Kemi. **16**, 215–219 (1960)
J. Lee, H.-J. Cho, B.-J. Jung, N.-S. Cho, H.-K. Shim, Macromolecules **37**, 8523–8529 (2004)
F. Li, D. Ramage, J.S. Lannin, J. Conceicao, Phys. Rev. B **44**, 13167–13170 (1991)
G. van Lier, C. van Alsenoy, V. van Doren, P. Geerlings, Chem. Phys. Lett. **326**, 180–185 (2000)
P.-O. Löwdin, Phys. Rev. **97**, 1474–1489 (1955)
M. Mattori, K. Mogi, Y. Sakai, and T. Isobe, J. Phys. Chem. A **104**, 10868–10872 (2000)
S. Maeda, Y. Harabuchi, Y. Ono, T. Takeguchi, K. Morokuma, Int. J. Quantum Chem. **115**, 258–269 (2015)
S. Maeda, K. Ohno, J. Chem. Phys. **124**(174306), 1–7 (2006)
T. Matsumoto, K. Tanaka, K. Tanaka, Y. Chujo, Dalton Trans. **44**, 8697–8707 (2015)
H. Mayr, M. Breugst, A.R. Ofial, Angew. Chem. Int. Ed. **50**, 6470–6505 (2011)
T. Miyahara, H. Nakatsuji, H. Sugiyama, J. Phys. Chem. A **117**, 42–55 (2013)
M. Montalti, A. Credi, L. Prodi, M.T. Gandolfi (eds.), *Handbook of Photochemistry*, 3rd edn. (CRC Press, Boca Raton, Fl, 2006), p. 619
K. Morokuma, Acc. Chem. Res. **10**, 294–300 (1977)
M. Moteki, S. Maeda, K. Ohno, Organometallics **28**, 2218–2224 (2009)
R.S. Mulliken, J. Chem. Phys. **2**, 782–793 (1934)
R.S. Mulliken, J. Chem. Phys. **23**, 1833–1840 (1952)
R.S. Mulliken, J. Chem. Phys. **23**, 1841–1846 (1955)

R.S. Mulliken, W.C. Ermler, *Diatomic Molecules: Results of Ab Initio Calculations* (Academic, New York, 1977), pp. 33–38
L.A. Nafie, T.B. Freedman, J. Chem. Phys. **78**, 7108–7115 (1983)
A. Nakahashi, Y. Yaguchi, N. Miura, M. Emura, K. Monde, J. Natl. Prod. **74**, 707–711 (2011)
K. Nakamae, T. Nishino, H. Ohkubo, J. Macromol. Sci. -Phys. **B30**, 1–23 (1991)
Y. Naruse, J. Ma, K. Takeuchi, T. Nohara, S. Inagaki, Tetrahedron **62**, 4491–4497 (2006)
T. Nishino, K. Nishi, T. Tendo, S. Tamba, A. Mori, K. Tanaka, D. Kawaguchi, K. Tanaka, *Structure, Mechanics, and Surface Properties of Poly(3-alkylthiophene), Report of the 144th Poval Committee, July 5* (Kyoto Univ, Japan, 2014)
K. Ohno, S. Maeda, Chem. Phys. Lett. **384**, 277–282 (2004)
T.N. Olney, N.M. Cann, G. Cooper, C.E. Brion, Chem. Phys. **223**, 59–98 (1997)
A.A. Ovchinnikov, Theoret. Chim. Acta **47**, 297–304 (1978)
R.G. Parr, R.G. Pearson, J. Am. Chem. Soc. **105**, 7512–7516 (1983)
L. Pauling, *The Nature of the Chemical Bond* (Cornell Univ. Press, Ithaca, N.Y., 1960)
R.G. Pearson, J. Am. Chem. Soc. **85**, 3533–3539 (1963)
R.G. Pearson, J. Chem. Educ. **45**, 581–586 and 643–648 (1968)
A. Peeters, C. van Alsenoy, F. Bartha, F. Bogár, M.–L. Zhang, V. van Doren, Int. J Quantum Chem. **91**, 32–38 (2003)
J.P. Perdew, R.G. Parr, M. Levy, J.L. Balduz Jr., Phys. Rev. Lett. **49**, 1691–1694 (1982)
A.E. Reed, F. Weinhold, J. Chem. Phys. **78**, 4066–4073 (1983)
A.E. Reed, R.B. Weinstock, F. Weinhold, J. Chem. Phys. **83**, 735–746 (1985)
A.E. Reed, F. Weinhold, J. Chem. Phys. **83**, 1736–1740 (1985)
A.E. Reed, L.A. Curtiss, F. Weinhold, Chem. Rev. **88**, 899–926 (1988)
J.R. Rumble (ed.), CRC Handbook of Chemistry and Physics, 99th edn. 2018–2019 (CRC Press, Boca Raton, Fl., 2018)
P. von Ragué Schleyer, C. Maerker, A. Dransfeld, H. Jiao, N.J.R. van Eikema Hommes, J. Am. Chem. Soc. **118**, 6317–6318 (1996)
J. Sauer, R. Sustmann, Angew. Chem. Int. Ed. Engl. **19**, 779–807 (1980)
H.B. Schlegel, WIREs Comput. Mol. Sci. **1**, 790–809 (2011)
C. Shang, Z.-P. Liu, J. Chem. Theory Comput. **8**, 2215–2222 (2012)
K. Shizu, M. Uejima, H. Nomura, T. Sato, K. Tanaka, H. Kaji, C. Adachi, Phys. Rev. Applied **3**(014001), 1–7 (2015a)
K. Shizu, H. Noda, H. Tanaka, M. Taneda, M. Uejima, T. Sato, K. Tanaka, H. Kaji, C. Adachi, J. Phys. Chem. C **119**, 26283–26289 (2015b)
O.V. Sizova, L.V. Skripnikov, A. Yu Sokolov, J. Mol. Struct. THOECHEM **870**, 1–9 (2008)
Z. Špitalský, T. Bleha, Macromol. Theory Simul. **10**, 833–841 (2001)
P.J. Stephens, J.J. Pan, F.J. Devlin, K. Krohn, T. Kurtán, J. Org. Chem. **72**, 3521–3536 (2007)
R. Stowasser, R. Hoffmann, J. Am. Chem. Soc. **121**, 3414–3420 (1999)
K. Tanaka, K. Ohzeki, S. Nankai, T. Yamabe, H. Shirakawa, J. Phys. Chem. Solids **44**, 1069–1075 (1983)
K. Tanaka, S. Wang, T. Yamabe, Synth. Metals **30**, 57–65 (1989)
K. Tashiro, Prog. Polym. Sci. **18**, 377–435 (1993)
R. Taylor, J.P. Hare, A.K. Abdul-Sada, H.W. Kroto, J. Chem. Soc. Chem. Commun. **1990**, 1423–1425 (1990)
D. Toptygin, J. Fluoresc. **13**, 201–219 (2003)
K.W. Törnroos, Acta Crystallogr. Sect. C **50**, 1646–1648 (1994)
E. Tsivion, S.P. Veccham, M. Head-Gordon, ChemPhysChem **18**, 184–188 (2017)
S.P. Verevkin, H.-D. Beckhaus, C. Rüchardt, R. Haag, S.I. Kozhushkov, T. Zywietz, A. de Meijere, H. Jiao, P. von R. Schleyer, J. Am. Chem. Soc. **120**, 11130–11135 (1998)
K.B. Wiberg, Tetrahedron **24**, 1083–1096 (1968)
K.B. Wiberg, Angew. Chem. Int. Ed. Engl. **25**, 312–322 (1986)
R.B. Woodward, R. Hoffmann, J. Am. Chem. Soc. **87**, 395–397 (1965)
R.B. Woodward, R. Hoffmann, Angew. Chem. Intern. Ed. Engl. **8**, 781–853 (1969)

S. Xiao, M. Nguyen, X. Gong, Y. Cao, H.B. Wu, D. Moses, A.J. Heeger, Adv. Funct. Mater. **13**, 25–29 (2003)
K. Yoshizawa, M. Hatanaka, A. Ito, K. Tanaka, T. Yamabe, Chem. Phys. Lett. **202**, 483–488 (1993)
C.-G. Zhan, J.A. Nichols, D.A. Dixon, J. Phys. Chem. A **107**, 4184–4195 (2003)
C.-G. Zhen, U. Becker, J. Kieffer, J. Phys. Chem. A **113**, 9707–9714 (2009)

Chapter 3
Fundamentals of the Analysis Tools

Abstract The concept of molecular orbitals (MO's) is of central importance in theoretical chemistry and is widely used by not only theoreticians but by experimentalists for understanding the electronic structures of molecules. In this chapter, simple bases or essential ingredients of the MO methods including the Hartree-Fock (HF), post-HF, density functional theory (DFT), and other useful calculation methods are to be described for convenience to effectively employ them. Although a few formulations are to appear in this chapter just for the assistance of understanding the outlook of the theories, those will be simplified as much as possible.

Keywords Molecular orbital · Hartree-Fock · Post-Hartree-Fock · Density functional · Crystal orbital

3.1 Molecular Orbital Calculations

The electronic structure of a molecule is best described by the Schrödinger equation based on quantum mechanics. Within the adiabatic (also often called Born-Oppenheimer, although there is a minute difference which can be neglected here) approximation and the non-relativistic scheme for simplicity, this equation is written as follows:

$$\widehat{H}\Phi_I = E_I\Phi_I \tag{3.1}$$

$$\widehat{H} = -\sum_{\mu}\frac{\nabla_{\mu}^2}{2} - \sum_{\mu,\mathrm{A}}\frac{Z_{\mathrm{A}}}{r_{\mu\mathrm{A}}} + \sum_{\mu>\nu}\frac{1}{r_{\mu\nu}} \tag{3.2}$$

where operator $\widehat{H}$ denotes the Hamiltonian and Φ_I and E_I are the exact wavefunction and energy, respectively, of the electrons in a molecule in the Ith state (see the Glossaries 1).

K. Tanaka, *Theoretical Chemistry for Experimental Chemists*,
https://doi.org/10.1007/978-981-15-7195-4_3

Glossaries 1: Some notations

Φ_I wavefunction of the Ith state ($I = 0$ represents the ground state)
E_I electronic energy of the Ith state
Z_A magnitude of the positive charge (e^+) of nucleus A
$r_{\mu A}$ distance between electron μ and nucleus A
$r_{\mu\nu}$ distance between electrons μ and ν
$\mathbf{r}$ coordinate of an arbitrary electron
r_A distance between an arbitrary electron and nucleus A
R_{AB} distance between nuclei A and B

It is well known that the Schrödinger equation constitutes an eigenvalue problem whose eigenfunction and eigenvalue correspond to wavefunction and energy, respectively. It is noted that the atomic unit listed in the Glossaries 2 is employed throughout this chapter for simplicity unless specially noted.

Glossaries 2: Atomic unit

Mass of a free electron

$$m_e = 1 \quad 9.1094 \times 10^{-31}\text{kg}$$

Bohr radius

$$a_0 = \frac{\hbar^2}{m_e e^2} = 1 \quad 5.2918 \times 10^{-11}\text{m}$$

Energy

$$E_h = \frac{m_e e^4}{\hbar^2} = 1 \quad 4.3597 \times 10^{-18}\text{J}$$

Time

$$\tau_0 = \frac{\hbar}{E_h} = \frac{\hbar^3}{m_e e^4} = 1 \quad 2.4189 \times 10^{-17}\text{s}$$

Elementary charge

$$+e = 1 \quad 1.6022 \times 10^{-19}\text{C}$$

Planck's constant

$$\hbar \equiv \frac{h}{2\pi} = 1 \quad 1.0546 \times 10^{-34}\text{J s}$$

Dielectric constant in vacuum

$$4\pi\varepsilon_0 = 1 \quad 1.1127 \times 10^{-10} \mathrm{C^2 J^{-1} m^{-1}}$$

The first and the second terms in the Hamiltonian $\widehat{H}$ in Eq. (3.2) are called, respectively, kinetic operator and nuclear attraction potential, and the third term is interelectron repulsion potential. It is the presence of this third term that makes it impossible to analytically solve the Schrödinger equation in case of molecules having multielectron as is well known. In this sense, one has to introduce a certain approximation to solve the Schrödinger equation. Theoretical chemists working for the description of atoms and molecules based on quantum mechanics have long been making efforts to improve this approximation as well as solid-state physicists dealing with solids.

The most popular way of approximation is to use the molecular orbitals (MO's) for a description of the electronic structure of molecules. Earlier MO methods include the Hückel (Hückel 1930) and the extended Hückel (Hoffmann 1963) schemes. The former is the simplest one dealing with only π electrons in π-conjugated organic molecules employing simplified parameters called Coulomb and resonance integrals. The latter can handle all the valence electrons including both σ and π electrons. Both of these methods are considered to be too simple to afford quantitative results but nonetheless provide certain insight particularly into the discussion of chemical reactions.

Currently, there are two major tools for the MO calculations: one is the HF method and another is based on the DFT method. The former is making further progress to the various post-HF methods, which give more accurate results compared with those by the ordinary HF method and nowadays are being selected toward obtaining more quantitative results particularly to the theoreticians. On the other hand, the latter has recently joined the theoretical chemistry method and seems to be more tractable due to less time-consuming and providing a certain accuracy. In the next few sections, these methods are to be elucidated.

3.1.1 Hartree-Fock (HF) Method

3.1.1.1 Basic Idea

The ordinary MO calculations in quantum chemistry have traditionally been performed with the use of the HF method described above. This has been and is still now considered to give reference to quantum chemical calculations for molecules and polymers. For instance, the HF scheme also gives the starting point to launch into the post-HF calculation which can more quantitatively describe the electronic structures. In this sense, comprehension of the basic concept of the HF method would be useful to the users of theoretical chemistry as well.

The HF framework lies in the usage of the wavefunction called "orbital" based on the one-electron approximation, i.e., an orbital is a function of the coordinate of one electron designated as **r**. This feature is completely different from that of the formal wavefunction Φ obtained from the Schrödinger equation in Eq. (3.1), since Φ is a function of the coordinates of all the electrons in the molecule.

In case of a molecule, the orbitals are called MO's as is well known. We set a definite function called basis set to represent each MO in terms of their linear combinations. The basis set functions originally ought to consist of atomic orbitals (AO's) for general multielectron atoms named by Slater-type orbitals (STO's) having the exponential-function type of exp(-ζr) inherent to each atom in the molecule, where r stands for the nucleus-electron distance and ζ is the orbital exponent representing the degree of expansion of the orbital. However, since there is some difficulty to use STO's in the actual calculation, they are usually further decomposed into a linear combination of Gaussian-type orbitals (GTO's) having the exponential-function type of $\exp(-\zeta r^2)$ with which the calculation becomes far more accessible. The MO is thus represented with the linear combination of the STO's as follows:

$$\psi_i(\mathbf{r}) = \sum_p c_{ip}\chi_p^{\mathrm{STO}}(\mathbf{r}) \tag{3.3}$$

where each χ_p^{STO} is further expanded with χ_q^{GTO} as follows:

$$\chi_p^{\mathrm{STO}}(\mathbf{r}) = \sum_q d_{pq}\chi_q^{\mathrm{GTO}}(\mathbf{r}) \tag{3.4}$$

the coefficients d_{pq} being pre-determined.

There are several ways of selection of the basis set to represent MO in accordance with the characteristics and expansion numbers of GTO's, so that one should be careful to this selection depending on one's problem. The details of the basis set will be summarized in Sect. 3.6. In any case, the coefficients of the basis set, e.g., c_{ip} in Eq. (3.3), to build up each MO are to be decided by the HF calculation. There is even a tendency to directly call the selected GTO's basis set.

3.1.1.2 Ground State

Within the framework of the HF method, the ground-state wavefunction Ψ_{HF} and the electronic energy E_{HF} of a molecule having N electrons are represented as follows:

$$\Phi(\mathbf{r}_1, \mathbf{r}_2, \cdots\cdots, \mathbf{r}_N) \simeq \Psi_{\mathrm{HF}} = \frac{1}{\sqrt{N!}}\det[\psi_1(\mathbf{r}_1)\psi_2(\mathbf{r}_2)\cdots\cdots\psi_N(\mathbf{r}_N)] \tag{3.5}$$

$$E_{\mathrm{HF}} = \sum_{i=1}^{N} H_i + \frac{1}{2}\sum_{i,j=1}^{N}(J_{ij} - K_{ij}) \tag{3.6}$$

where MO ψ_i of the ith level is set to also include the spin information of the electron accommodated therein, i.e., ψ_i is a spin-orbital (or spin MO). The right-hand side in Eq. (3.4) signifies the combination of MO's which satisfies the Pauli exclusion principle and is called Slater determinant. The total energy of a molecule is given by the sum of E_{HF} and the internuclear repulsion potential, i.e.,

$$E_{\mathrm{tot}} = E_{\mathrm{HF}} + \sum_{\mathrm{A}>\mathrm{B}} \frac{Z_{\mathrm{A}} Z_{\mathrm{B}}}{R_{\mathrm{AB}}} \tag{3.7}$$

Notations used are described in Glossaries 3.

Glossaries 3: Molecular integrals

H_i core integral

$$H_i = \int \psi_i^*(\mathbf{r}) \left[-\frac{1}{2}\nabla^2 - \sum_{\mathrm{A}} \frac{Z_{\mathrm{A}}}{r_{\mathrm{A}}} \right] \psi_i(\mathbf{r})\, \mathrm{d}\mathbf{r} \tag{3.8}$$

where all the variables are defined in Glossaries 1. Note again that only one electron is dealt with in the HF method.

J_{ij} Coulomb integral

$$J_{ij} = \int\int \frac{|\psi_i(\mathbf{r})|^2 |\psi_j(\mathbf{r}')|^2}{|\mathbf{r} - \mathbf{r}'|} \mathbf{drdr}' \tag{3.9}$$

K_{ij} exchange integral

$$K_{ij} = \int\int \frac{\psi_i^*(\mathbf{r})\psi_j^*(\mathbf{r}')\psi_i(\mathbf{r}')\psi_j(\mathbf{r})}{|\mathbf{r} - \mathbf{r}'|} \mathbf{drdr}' \tag{3.10}$$

The MO ψ_i is obtained by solving the following HF equation constructed for the concerning molecule instead of the exact Schrödinger equation (Roothaan 1951). The HF equation for a closed-shell molecule (N; even number) is essentially an eigenvalue problem for the following HF operator $\widehat{F}_{\mathrm{HF}}$. It is shown as follows:

$$\begin{aligned} \widehat{F}_{\mathrm{HF}} \psi_i(\mathbf{r}) &= \varepsilon_i \psi_i(\mathbf{r}) \\ \widehat{F}_{\mathrm{HF}} &\equiv \widehat{H}_{\mathrm{core}} + \sum_{i=1}^{N/2} \left(2\widehat{J}_i - \widehat{K}_i \right) \end{aligned} \tag{3.11}$$

the notations being listed in Glossaries 4.

Glossaries 4: Energy and operators

ε_i Energyε of the MO ψ_i (also called energy level or MO level)

$\widehat{H}_{\text{core}}$ Core Hamiltonian operator

$$\widehat{H} = -\frac{1}{2}\nabla^2 - \sum_{\mathrm{A}} \frac{Z_{\mathrm{A}}}{r_{\mathrm{A}}} \tag{3.12}$$

$\widehat{J}_i$ Coulomb repulsion operator

$$\widehat{J}_i = \int \frac{|\psi_i(\mathbf{r}')|^2}{|\mathbf{r} - \mathbf{r}'|} d\mathbf{r}' \tag{3.13}$$

$\widehat{K}_i$ Exchange repulsion operator

$$\widehat{K}_i = \psi_i(\mathbf{r}) \int \frac{\psi_i^*(\mathbf{r}')\widehat{P}_{\mathbf{r}\mathbf{r}'}}{|\mathbf{r} - \mathbf{r}'|} d\mathbf{r}' \tag{3.14}$$

$\widehat{P}_{\mathbf{r}\mathbf{r}'}$ Permutation operator to change the variable $\mathbf{r}$ to $\mathbf{r}'$.

In order to build the above operators $\widehat{J}_i$ and $\widehat{K}_i$, one needs the very MO's to enumerate the averaged potential, which means that one first has to assume a set of certain tentative MO's to start to solve the HF equation and then the resulting set of MO's obtained as the solution is used to build up the new averaged potential. The HF equation is thus iteratively solved until one eventually gets the stable set of MO's, which is called self-consistent field (SCF) solution or simply SCF-MO. The solution of the HF equation consists of the MO's ψ_i and their energies ε_i. Information of the solved MO's is explicitly represented by the actual values of coefficients of the basis set for each MO. Note there can also be obtained the MO's unoccupied with electrons which is called unoccupied or virtual MO's in addition to the occupied MO's in the ground state.

In order to express the total electron configuration of a molecule, one has to construct an antisymmetrized combination of the product of the MO's occupied with electrons to satisfy Pauli's exclusion principle. This combination eventually leads to a determinant referred to as a Slater determinant or single determinant in Eq. (3.5). A single determinant thus constructed with the occupied MO's gives a relatively simple wavefunction of the molecule in the ground state Ψ_0 describing a single configuration as shown in Fig. 3.1. Figure 3.1a shows an example of the ground state of a molecule with an even number of electrons and is often called the closed-shell structure, whereas Fig. 3.1b with odd numbers of electrons and is called the open-shell structure. There are two kinds of representations for the open-shell structure as shown in Figs. 3.1b, c. In the former, two electrons with α and β spins occupy the same spatial part of MO and, in the latter, those occupy the different spatial parts. In this sense, the former is called restricted open HF (ROHF) scheme (Roothaan 1960) and the latter is unrestricted HF (UHF) scheme (Pople and Nesbet 1954). The unrestricted model is also expressed as different orbitals for different spins (DODS). Although UHF method inevitably includes the spin contamination from the quantum

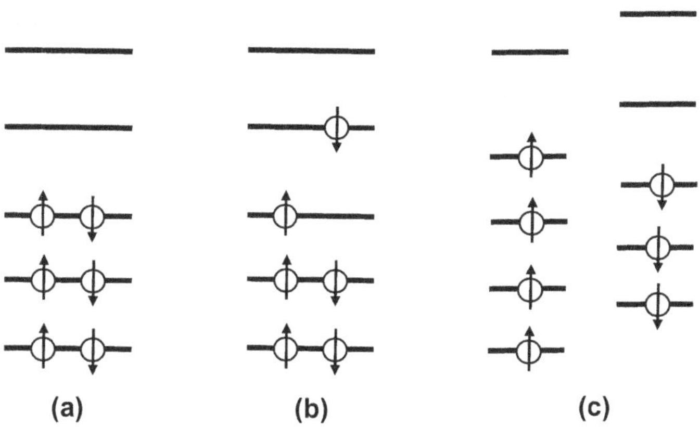

Fig. 3.1 Examples of MO energy levels obtained by each HF scheme. **a** Closed-shell structure, **b** open-shell structure (restricted open HF (ROHF) picture), and **c** open-shell structure (unrestricted open HF (UHF) picture). Small circles with the up and down arrows signify the electrons with α and β spins, respectively

mechanical points of view, it gives lower energy than ROHF due to larger freedom of variation. The spin contamination in the UHF method can be annihilated by a certain numerical procedure. On the other hand, there is no spin contamination in the ROHF description from the beginning. To the author's knowledge, the UHF tends to be more frequently employed for the open-shell structure calculation. The spin density values resulting from the unrestricted calculations are often used and has already been explained in Sect. 2.3.

Since the HF equation is for an electron moving in the averaged potential of the other electrons as mentioned above, it does not include the direct interelectron interactions. That is, the electron correlation is incompletely incorporated into the HF scheme and correlation energy E_{corr} virtually defined as follows:

$$E_{\mathrm{corr}} = E_{\mathrm{exact}} - E_{\mathrm{HF}} \tag{3.15}$$

is inevitably shown up, where E_{exact} stands for the electronic energy obtained from the exact Schrödinger equation and E_{HF} being ideally the best value of the electronic energy from the HF framework (Löwdin 1955; Sinanoğlu 1964). This best value is often called HF limit. The existence of correlation energy is a typical drawback in the HF method and inclusion of the electron correlation effect is desirable for more quantitative estimations of bond lengths, binding energy, excitation energy, and so on.

There are several simplified versions of the HF method being often called semiempirical HF ones that deal with only the valence electrons and use some effective parametrizations to build up molecular integrals shown in Glossaries 3 (Pople and Segal 1966; Del Bene and Jaffé 1969; Bingham et al. 1975). Although the semiempirical HF methods generally afford less quantitative results compared with that of the ordinary HF method, they are still sometimes employed for convenience due to quite short computation time required. On the other hand, the calculation scheme without any empirical parameters or abbreviation of molecular integrals is called non-empirical HF method (or simply HF method), which is employed in most cases.

3.1.1.3 Excited State

Electron excitations in the HF picture are represented by using the unoccupied (or virtual) MO's mentioned above as the destination of the excited electron. Various excited-state configurations due to one-electron excitation from the ith to the jth MO's are illustrated in Fig. 3.2. Care should be taken that the excitation energy $\Delta E_{i\to j}$ is not only the energy difference in the ith and the jth MO's but includes the adjustment occurring from the above J_{ij} and K_{ij} integrals in Eqs. (3.9) and (3.10). The one-electron excitation energies in the framework of the HF method is thus expressed as follows:

$$^{1}\Delta E_{i\to j} = \varepsilon_j - \varepsilon_i - J_{ij} + 2K_{ij} \tag{3.16}$$

for the singlet-state excitation (see Figs. 3.2a–c) and

$$^{3}\Delta E_{i\to j} = \varepsilon_j - \varepsilon_i - J_{ij} \tag{3.17}$$

for the triplet-state excitation (see Figs. 3.2d–f). It is obvious that the triplet-state excitation energy is smaller than the corresponding singlet-state one due to the lack of exchange integral K_{ij} having always a positive value. It is noted that, however, the triplet-state excitation is forbidden since the corresponding oscillator strength for this excitation (see Sect. 2.5) vanishes due to the integration of different spin states between the initial and the final configurations.

Formally, there can be excited-state configurations due to two-electron excitation as also shown in Fig. 3.3. However, one cannot accurately estimate the excitation

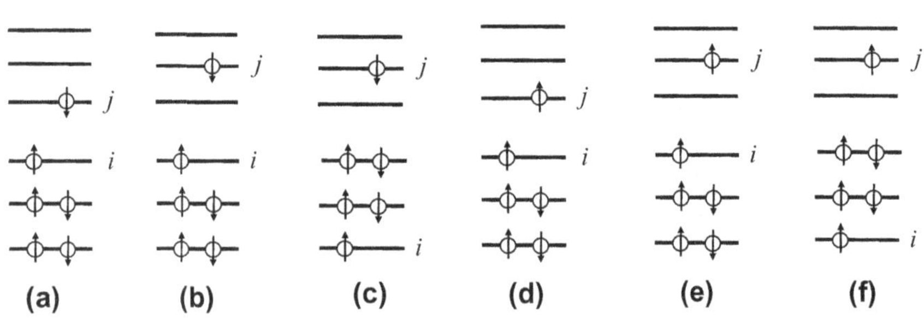

Fig. 3.2 **a–c** Various configurations of one-electron singlet excitation (singles) from the ith to the jth MO. Small circles with the up and down arrows signify the electrons with α and β spins, respectively. **d–f** represent the triplet excitation corresponding to (**a**)–(**c**)

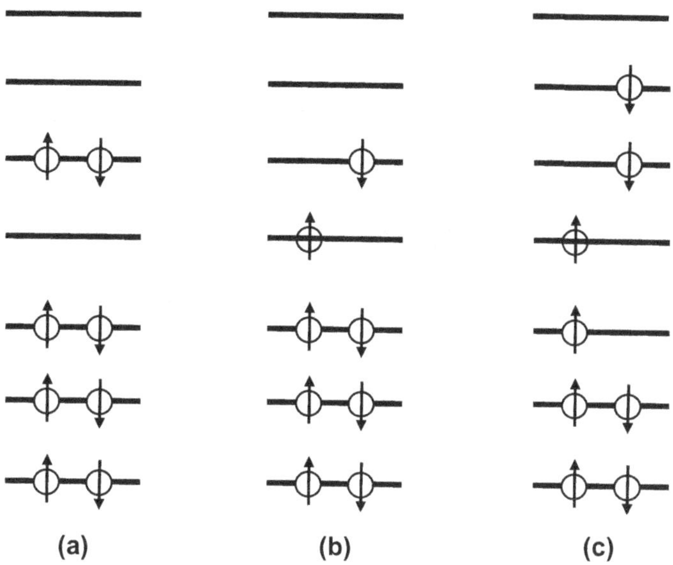

Fig. 3.3 Examples of multielectron-excitation configurations. **a** Two-electron excitation (doubles), **b** and **c** three-electron excitations (triples). Small circles with the up and down arrows signify the electrons with α and β spins, respectively

energy within the HF framework in principle since the description of the energy values of the unoccupied MO's, to which the electrons are supposed to be excited, is just additionally obtained to those of the occupied MO's and rather too simple due to the lack of electron correlation.

3.1.2 Post-HF Methods

Calculation methods taking the electron correlation into consideration more explicitly is known as the post-HF scheme. This can be used for (i) the improvement in the description of the ground state and (ii) the estimation of a more realistic value of the excitation energy. Several categories have been developed in this scheme as illustrated in Fig. 3.4.

A comparatively simple method to consider the electron correlation is the Møller-Plesset (MP) method in which the difference between the HF operator and the exact Hamiltonian is regarded as the perturbation (Møller and Plesset 1934). For instance, the second-order perturbation theory for the MP method is called MP2. Incorporation up to the higher order perturbation such as MP4 or more is also possible in recent years.

The basic idea in the post-HF method is to adopt the mixing of the excited-state configurations Ψ_I's with the ground-state one Ψ_0 represented by a single determinant

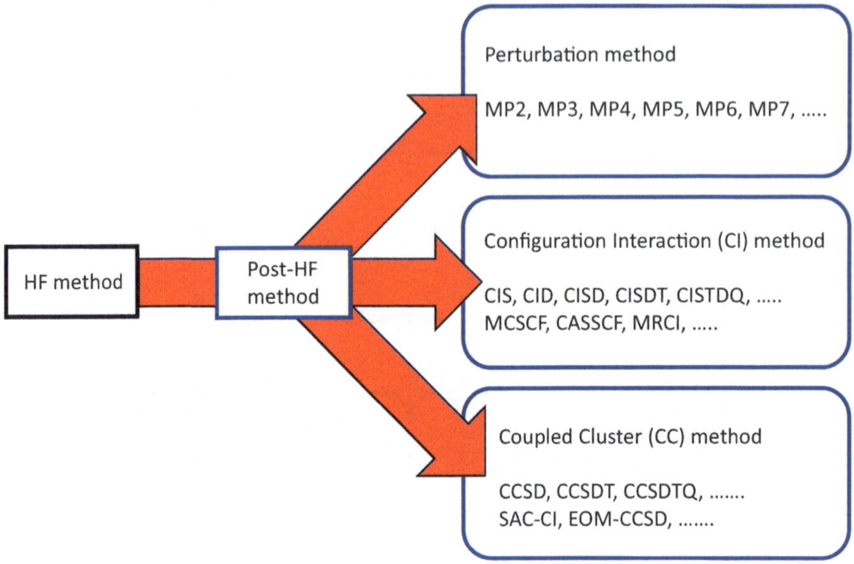

Fig. 3.4 Flow chart of HF and various post-HF methods. See the text as to abbreviated names

obtained by the HF method as is described by

$$\Phi_0 \simeq \Psi_0 + \sum_I c_I \Psi_I \tag{3.18}$$

in order to more effectively approximate the exact ground-state wavefunction Φ_0 in Eq. (3.1). In the configuration interaction (CI) method, the ratio of the mixing is determined with the use of the variation method (Condon 1930). Although the full CI method in which all the possible excitation configurations are employed gives the best result in the range of CI method, it is quite time-consuming and might not be practical. Hence, the restricted CI method is usually employed (see Fig. 3.4) including limited numbers of excitations are in consideration. The restricted CI method can be classified in terms of the number of excited electrons for building up the excitation configurations as sequentially listed in what follows:

Single excitations: CIS (CI singles)
Double excitations: CID or DCI (CI doubles)
Single + double excitations: CISD or SDCI (CI singles and doubles)
Single + double + triple excitations: CISDT (CI singles, doubles, and triples)
Single + double + triple + quadruple excitations: CISDTQ (CI singles, doubles, triples, and quadruples)

Actually, recent calculations of excitation energies often employ the CIS version as well as the time-dependent DFT (TD-DFT) version described in Sect. 3.2.

It is noted that these restricted CI considerations do not satisfy size consistency, i.e.,

$$E_{\mathrm{CI}}(M) \neq ME_{\mathrm{CI}}(1) \tag{3.19}$$

where $E_{\mathrm{CI}}(M)$ signifies the energy obtained for infinitely separated M-mers (Čížek 1966). The coupled-cluster (CC) method based on the selection of excitation configurations satisfying the size consistency has been developed for a reasonable approach. Like the restricted CI method, there is also a similar classification in the CC method such as CCSD, CCSDT, and so on.

For a molecule which has very small excitation energy, the ground-state configuration is energetically almost degenerate with that excitation configuration(s). In such a case, one ought to start from these mixed configurations state called multi-configuration. In this process, both the MO coefficients and the CI coefficients are simultaneously decided in the SCF process. This method is called the multiconfigurational SCF (MCSCF) approach (Frenkel 1934). Sometimes selection of the MO's near the HOMO and the LUMO as shown in Fig. 3.5 is denoted as complete active space (CAS). When all the possible excitations within this active space are included in the MCSCF method, it is called CASSCF method (Roos et al. 1980). It is noted that consideration of the electron correlation in MCSCF and CASSCF methods is still not complete, and further inclusion of the excited configurations in addition to multiconfigurations are to be considered, which is called multireference CI (MRCI) method (Whitten and Hackmeyer 1969). It is mentioned that MRCI method includes dynamic electron correlation and MCSCF includes nondynamic electron correlation. There are also schemes including dynamic electron correlation such as multireference SDCI (MR-SDCI), multireference MP2 (MR-MP2), CAS second-order perturbation theory (CASPT2), and so on. There is also a rather special post-HF method called

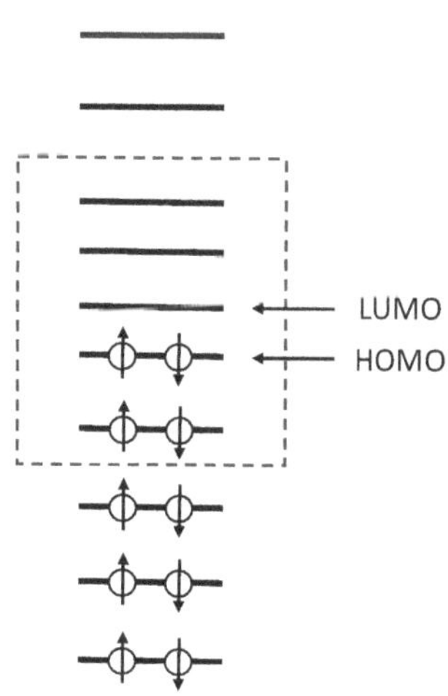

Fig. 3.5 Example of active space (surrounded by a dashed line). The size of CAS is expressed by (4, 5) in this example, where 4 and 5 signify the number of electrons and MO's, respectively

generalized valence bond (GVB) scheme (Goddard and Harding 1978) essentially akin to CASSCF procedure.

More sophisticated CI and CC methods are also applicable to the improvement of excitation energy. The CC method along with this line contains the symmetry-adapted cluster (SAC)-CI method which effectively shortens the computation time (Nakatsuji and Hirao 1978). Hence, when one requires more quantitative values of excitation energies, it is encouraged to consider these elaborate calculations.

3.2 Density Functional Theory (DFT) Calculations

3.2.1 Ground-State DFT

The DFT method was first developed in the physics field particularly for dealing with the inorganic or metal crystals and is often called the first-principles calculation by physicists. This method has recently been much more refined to achieve chemical accuracy when employed for the molecular systems. Therefore, it is now quite often used in the chemistry field as well. The calculation method based on the DFT was introduced into the category of theoretical chemistry after the 1990s since it can provide plausible results with reasonably short computation time in contrast with the HF method. In this section, a simple outlook of the DFT framework is to be given.

The theoretical background of DFT was established in the middle 1960s by the Hohenberg-Kohn theorem (Hohenberg and Kohn 1964) claiming that there is a one-to-one correspondence between the external potential $v(\mathbf{r})$ and the electron density $\rho(\mathbf{r})$ in the interactive electronic system. It is noted that the above external potential signifies the summation of the nucleus-electron interactions in atoms or molecules described by the second term in Eq. (3.2) and the potential due to the external electric, magnetic, or electromagnetic field, if any. The Hohenberg-Kohn theorem also claims that one can obtain the exact energy E of the Schrödinger equation if the exact $\rho(\mathbf{r})$ were to be employed. Hence, the concept of the DFT method starts from the electron density for obtaining its functional in principle. This concept rather differs from that of the conventional MO scheme including the HF method which starts from the "actual" wavefunction.

Hence, the energy of the concerning electronic system in the DFT scheme is described by functional of the electron density $\rho(\mathbf{r})$, i.e.,

$$E = E[\rho] = T[\rho] + U[\rho] = T_0[\rho] + \int \rho(\mathbf{r})v(\mathbf{r})d\mathbf{r} + \frac{1}{2}\int\int \frac{\rho(\mathbf{r})\rho(\mathbf{r}')}{|\mathbf{r}-\mathbf{r}'|}d\mathbf{r}d\mathbf{r}' + E_{XC}[\rho] \tag{3.20}$$

the bracket signifying the functional. T and U denote the kinetic and the interaction energies, respectively, in the concerning electronic system. In other words, T

corresponds to the first term of the Hamiltonian in Eq. (3.2) and U to the summation of the second and the third terms. T_0 in the rightmost side of Eq. (3.20) is made to represent the fictitious kinetic energy in the "non-interacting" electronic system having the density ($\mathbf{r}$), the third term the corresponding (classical) Coulomb energy, and $E_{XC}[\rho(\mathbf{r})]$ has a role of compensation as expressed by

$$E_{XC}[\rho] = T[\rho] - T_0[\rho] + U[\rho] - \frac{1}{2}\int\int \frac{\rho(\mathbf{r})\rho(\mathbf{r}')}{|\mathbf{r}-\mathbf{r}'|}\mathrm{d}\mathbf{r}\mathrm{d}\mathbf{r}' \tag{3.21}$$

but the most important implication of $E_{XC}[\rho(\mathbf{r})]$ is that it contains the exchange and the correlation energies, which is indicated by the subscript XC and is described in detail below.

It should be noted, however, that the Hohenberg-Kohn theorem affords prescription of obtaining neither actual electron density nor energy per se, since it is a kind of existence theorem. So one has to manage to perform the actual calculation formally based on Eq. (3.20). To express the functional, it has been proposed to employ "fictitious" spin-orbital set $\{\psi_i(\mathbf{r})\}$ for obtaining the electron density by

$$\rho(\mathbf{r}) = \sum_i^N \psi_i^*(\mathbf{r})\psi_i(\mathbf{r}) \tag{3.22}$$

with N being the number of electrons in the concerning molecule (Kohn and Sham 1965). This $\psi_i(\mathbf{r})$ is named Kohn-Sham (KS) orbital and each $\psi_i(\mathbf{r})$ accommodates one electron of either α or β spin. In this sense, the KS orbital is based on the "unrestricted" picture. The KS orbitals are used to make up a single determinant or KS determinant as follows:

$$\Psi_{KS} = \frac{1}{\sqrt{N!}}\det[\psi_1(\mathbf{r}_1)\psi_2(\mathbf{r}_2)\cdots\cdots\psi_N(\mathbf{r}_N)] \tag{3.23}$$

Similarly to the Slater determinant in the HF scheme (Eq. (3.5)) to satisfy the Pauli exclusion principle as well. Using these $\rho(\mathbf{r})$ and Ψ_{KS}, the energy E in Eq. (3.20) can be rewritten as follows:

$$E = \sum_\mu \left\{\int \Psi_{KS}^*\left(-\frac{\nabla_\mu^2}{2}\right)\Psi_{KS}\mathrm{d}\mathbf{r}\right\} + \frac{1}{2}\int\int \frac{\rho(\mathbf{r})\rho(\mathbf{r}')}{|\mathbf{r}-\mathbf{r}'|}\mathrm{d}\mathbf{r}\mathrm{d}\mathbf{r}' + \sum_{\mu,A}\left\{\int\left(-\frac{Z_A}{r_{\mu A}}\right)\rho(\mathbf{r})\mathrm{d}\mathbf{r}\right\} + E_{XC}[\rho(\mathbf{r})] \tag{3.24}$$

where $E_{XC}[\rho(\mathbf{r})]$ is again put for compensation in the KS framework with the use of the exact wavefunction Φ_{Exact}, being explicitly expressed as follows:

$$E_{\mathrm{XC}}[\rho(\mathbf{r})] \equiv \sum_{\mu}\left\{\int \Phi^{*}_{\mathrm{Exact}}\left(-\frac{\nabla^{2}_{\mu}}{2}\right)\Phi_{\mathrm{Exact}}\mathrm{d}\mathbf{r}_{\mu}\right\} - \sum_{\mu}\left\{\int \Psi^{*}_{\mathrm{KS}}\left(-\frac{\nabla^{2}_{\mu}}{2}\right)\Psi_{\mathrm{KS}}\mathrm{d}\mathbf{r}_{\mu}\right\}$$
$$+\sum_{\mu>\nu}\left\{\int\int \Phi^{*}_{\mathrm{Exact}}\left(\frac{1}{r_{\mu\nu}}\right)\Phi_{\mathrm{Exact}}\mathrm{d}\mathbf{r}_{\mu}\mathrm{d}\mathbf{r}_{\nu}\right\} - \frac{1}{2}\int\int \frac{\rho(\mathbf{r})\rho(\mathbf{r}')}{|\mathbf{r}-\mathbf{r}'|}\mathrm{d}\mathbf{r}\mathrm{d}\mathbf{r}' \tag{3.25}$$

The variables in Eqs. (3.24) and (3.25) are defined in Eq. (3.2) and Glossaries 1 in Sect. 3.1.

The quantity $E_{\mathrm{XC}}[\rho(\mathbf{r})]$ is called the exchange-correlation energy and consists of the following:

(i) difference between the kinetic energy based on the exact wavefunction Φ_{Exact} and that on the KS determinant Ψ_{KS} constructed from the KS orbitals as shown in Eq. (3.23) and
(ii) the exact interelectron exchange and correlation energies.

Although Eq. (3.20) seems straightforward and simple, rigorous estimation of $E_{\mathrm{XC}}[\rho(\mathbf{r})]$ in Eq. (3.25) contains various ordeals and provides challenging subjects.

Minimization process of the energy E in Eq. (3.24) eventually leads to the KS equation

$$\hat{F}_{\mathrm{KS}}\psi_i(\mathbf{r}) = \varepsilon_i\psi_i(\mathbf{r}) \tag{3.26}$$

$$\hat{F}_{\mathrm{KS}} \equiv \hat{H}_{\mathrm{core}} + \sum_{i=1}^{N}\hat{J}_i + V_{\mathrm{XC}}(\mathbf{r}) \tag{3.27}$$

where $\hat{F}_{\mathrm{KS}}$ is called the KS operator, and $\hat{H}_{\mathrm{core}}$ and $\hat{J}_i$ similarly defined in Glossaries 4 in Sect. 3.1. The new quantity $V_{\mathrm{XC}}(\mathbf{r})$ is the exchange-correlation potential defined by the functional differential,

$$V_{\mathrm{XC}}(\mathbf{r}) = \frac{\delta E_{\mathrm{XC}}[\rho(\mathbf{r})]}{\delta\rho(\mathbf{r})} \tag{3.28}$$

presence of which makes it different from the HF operator in Eq. (3.11).

The difference in the HF and the KS equations lies in that the exchange repulsion operator $\hat{K}_i$ is not included in the latter and instead the exchange-correlation potential $V_{\mathrm{XC}}(\mathbf{r})$ appears as in Eq. (3.26). This means that one could obtain the "exact" energy of the Schrödinger equation E_{Exact} including the correlation energy by properly solving the KS equation provided that the exact $V_{\mathrm{XC}}(\mathbf{r})$ is known. The KS equation if obtained is a one-electron equation like the HF equation and hence should be solvable with an appropriate procedure. In this sense, it is understood that the KS equation takes almost the same or even less time compared with that required for the HF equation to be solved. It is actually estimated that the computation time is proportional to N^3 for the KS equation to solve, whereas N^4 for the HF equation with N standing for the number of electrons.

Since the KS operator itself contains the electron density derived from the KS orbitals as in Eq. (3.22), the KS equation has to be iteratively solved starting from a certain assumption of the initial set of the KS orbitals until obtaining the SCF solution, i.e., the 'stable' set of KS orbitals as in the UHF equation, since the KS orbitals distinguish the α and β the spins. In order to represent the KS orbitals are mostly employed the Gauss-type orbitals as the basis sets like in the HF MO's. $E_{XC}[\rho(\mathbf{r})]$ and $V_{XC}(\mathbf{r})$ normally contain arithmetically complicated functions and hence require numerical integrations, for which the sample points called grids are generated in a precise manner around each atomic nucleus constructing the molecule to be examined. This allows us to calculate the precise total energy as well as the energy gradient.

It is of main importance to properly estimate the exchange-correlation potential $V_{XC}(\mathbf{r})$ or, in other words, exchange-correlation functional of good quality. This comes from that the results out of the KS equation are determined by the adaptability and the range of application of the exchange-correlation functional. Unfortunately, no systematic ways have been manipulated to enhance the approximation level of the functional and, hence, rather semiempirical efforts have been accumulated by referring to the 'more accurate' calculations, the experimental results, and so on. Along with this concept, quite a few functionals have been proposed, some of them being listed in Fig. 3.6 (Parr and Yang 1989; Cohen et al. 2012). The stage of developments of the DFT functionals are as follows:

(1) The simplest DFT functional is expressed only by the electron density such as $A[\rho(\mathbf{r})]$. This scheme is called local density approximation (LDA), in which, for instance, the Xα exchange functional proposed in early days (Slater 1951) and the VWN correlation functional (Vosko et al. 1980) are employed.

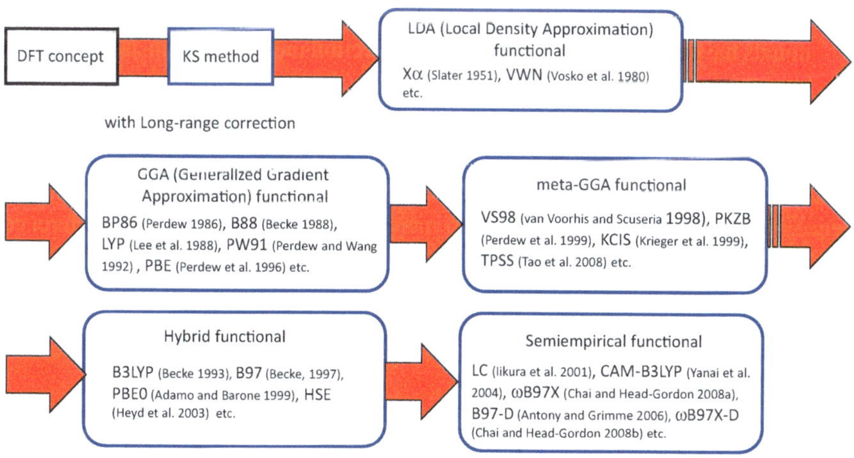

Fig. 3.6 Flow chart of the various functionals used in the DFT method

It is noted that most of the functionals used in the DFT method are named taken from the initials of the developers like VWN, LYP (see below), and so on. The namings are sometimes accompanied by the year number of the development like B88 (also see below).

(2) A sophisticated way of dealing with functionals is to include the gradient of density, $\nabla\rho(\mathbf{r})$, in the functional to correct the behavior of the density $\rho(\mathbf{r})$ (generalized gradient approximation: GGA). The GGA concept is employed in the BP86 (Perdew 1986), B88 (Becke 1988), LYP (Lee et al. 1988), PW91 (Perdew and Wang 1992), PBE (Perdew et al. 1996) functionals, and so on.

(3) Further refinement of the GGA functional includes the kinetic energy density τ and the higher order derivative of the density, $\nabla^2\rho(\mathbf{r})$. This procedure leads to the meta-GGA functionals such as VS98 (van Voorhis and Scuseria 1998), PKZB (Perdew et al. 1999), KCIS (Krieger et al. 1999), and TPSS (Tao et al. 2008).

(4) The inclusion of contribution from the exchange energy based on the HF method in the functional with a certain ratio is often adopted to make up, what is called, hybrid exchange functional. This incldes B97 (Becke 1997), PBE0 (Adamo and Barone 1999) and HSE (Heyd et al. 2003). One of the most often used hybrid functionals is the B3LYP consisting of a linear combination of three-parameter hybrid exchange functional B3 (Becke 1993) and the GGA-based correlation functional LYP with the coefficients decided by considering the experimental values of miscellaneous molecules as a benchmark (Curtiss et al 1991). Also, it is noted that self-interaction error (SIE) owing to the overcount of the Coulomb interaction of an electron itself is removed in the LYP-type functional.

(5) It is noted that the DFT method is not necessarily appropriate to describe weak interactions such as hydrogen bond and long-range interactions involving van der Waals interactions due to a lack of the exchange interactions between the long-separated electrons. To effectively deal with those within the DFT method, several frameworks such as the long-range correction (LC)-DFT (Iikura et al. 2001), Coulomb attenuating method (CAM)-B3LYP (Yanai et al. 2004), and ωB97X (long-range correction to B97) (Chai and Head-Gordon 2008a) functionals have been proposed. Moreover, correction including the dispersion interaction becomes possible by using B97-D (Antony and Grimme 2006) and ωB97X-D (Chai and Head-Gordon 2008b).

Rigorously speaking, the KS orbitals have no physical meaning by themselves, in principle, and hence they are different from the MO's obtained by the HF method. Hence, the KS determinant is alluded to merely the wavefunction in the "reference system without interaction". Nonetheless, both the KS orbitals and the HF MO's seem rather similar as seen in Fig. 3.7 (Stowasser and Hoffmann 1999), for instance, and such coincidence frequently happens. Therefore, these days, the KS orbitals are often called MO's as well but this naming ought to be considered with certain reservations due to the above reason.

Some features of the DFT method are listed in Table 3.1 in comparison with those of the HF method. It is noted that incorporation of the exchange and correlation effects into the DFT functionals with good balance is the most important to describe

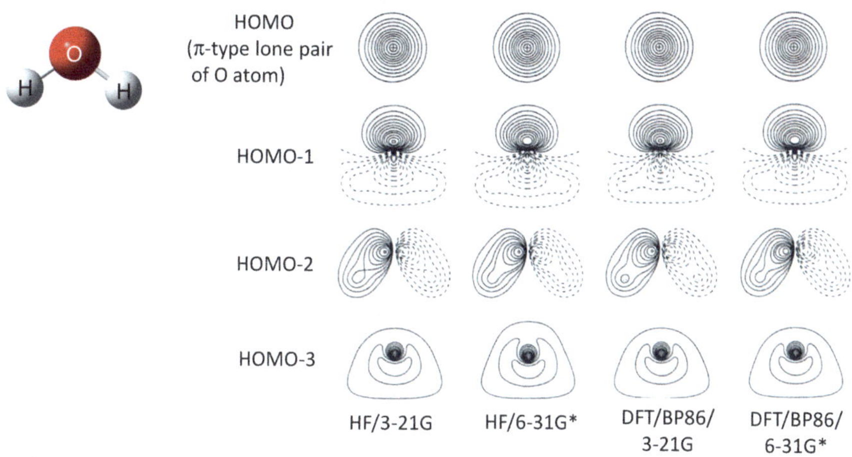

Fig. 3.7 MO patterns of a water molecule represented by contours with different calculation methods and basis sets. Adapted with permission from Stowasser and Hoffmann (1999). Copyright 1999 American Chemical Society

Table 3.1 Comparison of the HF and the DFT features

Items	HF	DFT
Geometry optimization	Good for molecules with a strong covalent bond	Good for molecules including metallic atoms
Binding energy	Underestimate	Overestimate
Atomic net charge and bond order	Dependent on basis set and on individual calc. method	Dependent on basis set and on individual calc. method
Spin state	Stabilization of high-spin state	Stabilization of low-spin state (hybrid functional stabilizes high-spin state due to resemblance to HF)
HOMO-LUMO gap	Overestimation	Underestimation
Excitation energy	Usually calculated by CIS	Calculated by TD-DFT
Activation Energy of the transition state (TS)	Tendency of overestimation	Tendency of underestimation
General Features	No electron correlation and hence requires the post-HF calculation. Complete exchange within the HF density	Incomplete electron exchange but includes partial electron correlation
Computation time	Moderate	Reasonably fast

molecular structures and molecular properties in successful manners with chemical accuracy.

3.2.2 Excited State in DFT

In order to obtain the excitation energies in the DFT method it is necessary to include an additional technique based on the TD-DFT theorem (Runge and Gross 1984). Hereinafter, a simple outline of the TD-DFT process is to be introduced. This theorem claims that time-dependent external potential $V_{\text{ext}}(\mathbf{r}, t)$ has a one-to-one correspondence to the time-dependent electron density $\rho(\mathbf{r}, t)$ leading to the time-dependent KS (TD-KS) equation

$$\left\{-\frac{1}{2}\nabla^2 + V_{\text{eff}}[\mathbf{r}, t; \rho(\mathbf{r}, t)]\right\}\psi_i(\mathbf{r}, t) = i\frac{\partial}{\partial t}\psi_i(\mathbf{r}, t) \tag{3.29}$$

where

$$V_{\text{eff}}[\mathbf{r}, t; \rho(\mathbf{r}, t)] = V_{\text{ext}}(\mathbf{r}, t) + \int \frac{\rho(\mathbf{r}', t)}{|\mathbf{r} - \mathbf{r}'|}d\mathbf{r} + \frac{\delta A_{\text{XC}}[\rho(\mathbf{r}, t)]}{\delta\rho(\mathbf{r}, t)} \tag{3.30}$$

and $A_{\text{XC}}[\rho(\mathbf{r}, t)]$ stands for unknown exchange-correlation functional depending on time. The TD-KS equation affords the output of dynamics of the electron system including the correlation effect. The time-independent exchange-correlation potential $V_{\text{XC}}[\rho(\mathbf{r})_t]$, with density $\rho(\mathbf{r})_t$ at fixed time t, is usually used instead of $\frac{\delta A_{\text{XC}}[\rho(\mathbf{r},t)]}{\delta\rho(\mathbf{r},t)}$ in Eq. (3.30), which is called adiabatic approximation. In this approximation, the word "adiabatic" implies that there is no retardation effect due to the instantaneous response of the electron density. This is particularly plausible for low-lying excited states derived from definite valence configuration.

Calculation of excitation energy is based on the combination of Runge-Gross theorem and dynamical response theory to the periodically oscillating external field, where the excitation energy is given as the poles of the dynamic polarizability (Jamorski et al. 1996). In the TD-DFT scheme, the one-electron excitation energy of either singlet or triplet excited state is accompanied by the combination of several transitions representing excitation configurations with different coefficients, which allows eventually to include the effect of relaxation of the MO's. Typical examples of the calculation of the excitation energies by the TD-DFT method have already been given in Sect. 2.5.

3.3 Crystal Orbital (CO) Calculations

3.3.1 Basic Idea

For regular one-dimensional (1D) polymers, quantum chemical treatment is also possible by the extension of the ordinary MO calculations. The 1D polymer treated here is defined to have the structure illustrated in Fig. 3.8a with infinite repetition of the unit cell indicated by the translation length. In this sense, 1D polymer essentially has the structure of 1D crystal with the alignment of an appropriate unit cell along one direction. Thus the 1D polymer is regarded as a simpler version of the ordinary 3D crystal. Some 1D polymers have the screw-axis symmetry defined by the combination of translation length (pitch length) and rotational angle around the screw-axis (Fig. 3.8b). DNA, α-helix of protein, carbon nanotube of helical type, or other organic helical polymers possess this pattern of symmetry. Examples of actual unit cells including those of 2D polymers are given in Fig. 3.9. Some calculation results of 1D polymers have already been described in Sect. 3.2.1, 2.6, and 2.8 antecedent to this Section.

Although it may seem difficult to deal with the infinite system, it is rather easy to handle from a mathematical point of view by the introduction of periodic boundary condition (also called Born–von Karman boundary condition) shown in Fig. 3.10. The ring shown here has the infinite diameter and hence an infinite number of the unit cells N ($N \to \infty$) so that the alignment of the unit cell can actually be regarded as a linear 1D polymer with infinite curvature. The final unit cell thus becomes the same as the first cell due to this condition. The condition mentioned above makes it possible to perform computation of the polymers and crystals with idealized structures. The

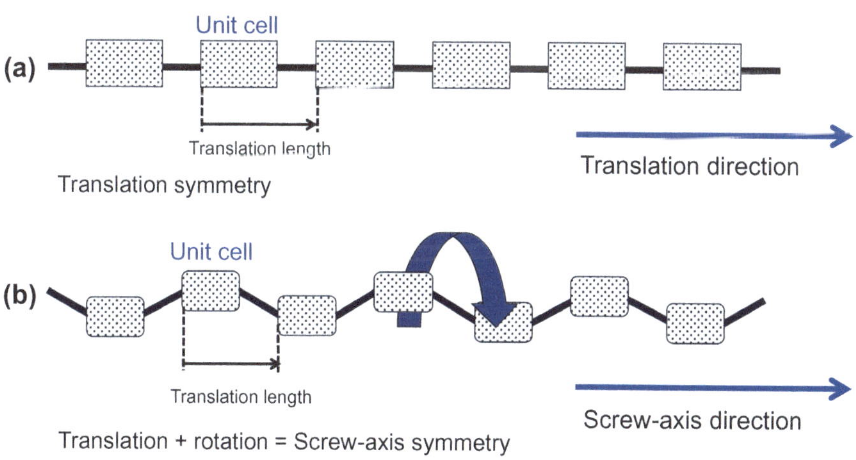

Fig. 3.8 Schematic structure of 1D polymers with **a** translation symmetry and with **b** screw-axis symmetry of pitch angle of 180°

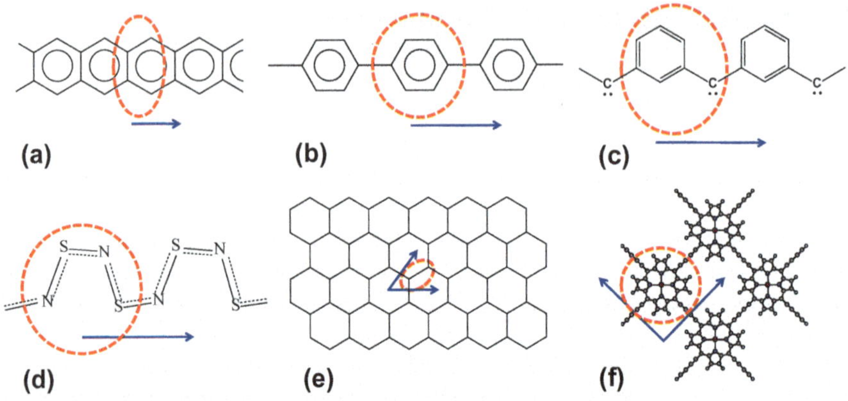

Fig. 3.9 Various 1D and 2D polymers: **a** polyacene, **b** poly(*p*-phenylene), **c** poly(*m*-phenylcarbene), **d** polythiazyl, **e** graphene, and **f** 2D porphyrin. Unit cells are shown by broken curves and translation length by arrows

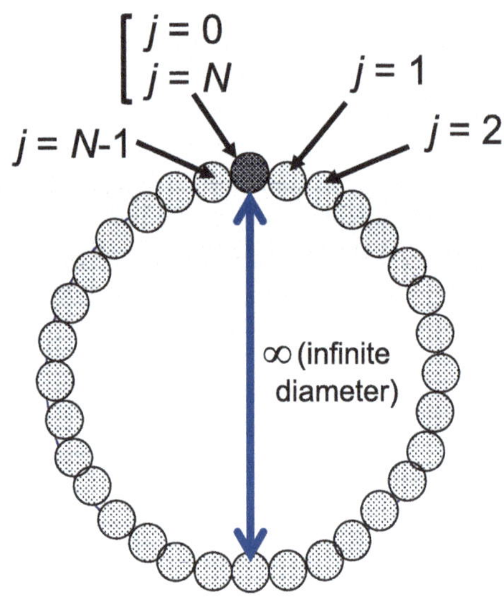

Fig. 3.10 Mathematical manipulation to treat 1D polymer with infinite length. Small circles represent the unit cells totally making a large ring with infinite diameter indicated by an arrow. The Nth cell ($j = N$) becomes the 0th cell ($j = 0$) shown in red due to this periodic boundary condition

periodic boundary condition brings about a sort of new quantum number called "wave vector" attached to the wavefunction and its energy to give, what is called, band structure. This is a usual and typical feature for a description of crystals including 1D polymers and is to be elucidated below.

3.3.2 *Wavefunction of 1D Polymer*

The wavefunction $\psi(x)$ describing 1D polymer can be translated by the translation length a and the translated one, $\psi(x+a)$, is "essentially the same" with the original wavefunction except for the phase factor λ being a complex number having the absolute value of unity. This relationship is represented by

$$\psi(x+a) = \lambda\psi(x) \tag{3.31}$$

Then for the successive translation one obtains

$$\left.\begin{aligned} \psi(x+2a) &= \lambda^2\psi(x) \\ \psi(x+3a) &= \lambda^3\psi(x) \\ \psi(x+4a) &= \lambda^4\psi(x) \\ &\vdots \\ \psi(x+Na) &= \lambda^N\psi(x) = \psi(x) \end{aligned}\right\} \tag{3.32}$$

The last equation in the above comes from the very periodic boundary condition shown in Fig. 3.10, which implies

$$\lambda^N = 1 \tag{3.33}$$

This equation is the usual circle equation giving λ as follows:

$$\lambda = \exp\left[\frac{2\pi iq}{N}\right] \quad (q = 0, 1, 2, \ldots\ldots, N-1) \tag{3.34}$$

where N is formally infinite and i imaginary unit. By setting

$$\frac{2\pi q}{Na} \equiv k \tag{3.35}$$

one can write λ more simply as follows:

$$\lambda = \exp[ika] \tag{3.36}$$

Hence, Eq. (3.31) is changed into

$$\psi(x+a) = \exp[ika]\psi(x) \tag{3.37}$$

where k is the reciprocal translation vector, often called wave vector or wavenumber, having the reverse dimension of the translation length in the range

$$0 \leq k < \frac{2\pi}{a} \quad \text{or} \quad -\frac{\pi}{a} \leq k < \frac{\pi}{a} \tag{3.38}$$

This range is called the first Brillouin zone or simply the Brillouin zone, in which k is continuously distributed due to infinitesimally small separation between each k value. The wavefunction of the form in Eq. (3.37) extends to all over the 1D polymer and is often referred to as the Bloch function since it satisfies the Bloch theorem concerning the translation of the wavefunction (Kittel et al. 2004), which is omitted here for simplicity. The physical meaning of k (in terms of $\hbar k$) expresses the momentum of electron accommodated in the CO.

We often employ the AO's for the actual description of the CO as the basis set, i.e.,

$$\psi_s(k,x) = \frac{1}{\sqrt{N}} \sum_{j}^{N} \sum_{\mu}^{\mathrm{AO}} \exp[ijka] C_{\mu,s}(k) \chi_\mu(x - ja) \tag{3.39}$$

(Del Re et al. 1967). This type of expression is often called the tight-binding approximation by the solid-state physicist since it perhaps seems for them an approximation to utilize the linear combination of AO's being "tightly bound" to the nuclei.

The glossaries for the CO are bit complicated than those for the MO and hence listed as follows:

Glossaries 5: Some notations for CO calculation

N	total number of the unit cells being actually infinite.
j	numbering of the unit cell is in the range from 0 to $N-1$ (see Fig. 3.10).
s	energy level of the CO ψ corresponding to, e.g., the energy level i of the MO ψ.
k	wave vector as mentioned above and assigned to each CO.
$\psi_s(k, x)$	CO with the energy level s and the wave vector k.
$\varepsilon_s(k)$	orbital energy of the CO $\psi_s(k, x)$.
χ_μ	μ-th AO in the unit cell.
$C_{\mu,s}(k)$	coefficient of χ_μ in the s-th CO described by the wave vector k.
a	translation length of the unit cell.

It is noted that both k and a should be vectors for the 2D or 3D crystal, but are scalars (x components) for the 1D polymer dealt with here. In this connection, the spatial coordinate in the CO's and the AO's is to be expressed by x instead of $\mathbf{r}$ for simplicity unless specially noted.

3.3.3 Details of the CO Calculation

Analysis in terms of CO's seems to be less popular among the experimental chemists compared with that by the MO's. Indeed, the calculation routine for the 1D polymer fabricated in the popular softwares mainly for the MO calculations is not necessarily complete, which one should bear in mind. But information obtained from the CO calculation is useful when one has to examine the electronic structure of a considerably long polymer. Although the CO calculation deals with an idealistic polymer with infinite length and regular repetition of the unit cells, information therefrom can afford a sort of necessary conditions for the appearance of the electronic properties of actual polymers.

The CO having the form of Eq. (3.39) is a kind of extended version of the ordinary MO modified with an introduction of the Bloch theorem necessary for the description of the 1D polymer. In this sense, the CO's are within the range of one-electron approximation and hence the HF or the DFT framework is still valid to combine with the CO calculation. Hence, the classification of basis set used for the CO is common to those for the MO. The CO calculation codes for 1D, 2D, and 3D crystals in terms of both the HF and the DFT methods are usually also implemented in most of the commercial MO calculation softwares. In such CO routines, the typical default numbers of k points in the Brillouin zone are, for instance, 100 for the 1D polymer and, hence, 100^2 and 100^3 for the 2D and the 3D crystals, respectively. Hence, it takes much more time to calculate these systems compared with the ordinary MO calculation with the same basis set.

Since it is rather boring to write down all the details of the calculation scheme of the 1D polymer (Del Re et al. 1967; Imamura and Fujita 1974), only the essential outline of that is to be described in the following. For instance, within the HF framework, the Fock equations for each k point considered in the Brillouin zone are constructed using the total density matrix $P_{\mu\nu}$ and solved to obtain the density matrix $R_{\mu\nu}(k)$ at that k point. The total density matrix between the central ($j = 0$) and the jth cells is obtained from integration of the density matrix at each k over the Brillouin zone in formal as follows:

$$P^j_{\mu\nu} = \frac{Na}{2\pi} \int_{-\frac{\pi}{a}}^{\frac{\pi}{a}} \exp[ijka] R_{\mu\nu}(k)\, dk \tag{3.40}$$

where the superscript j implies the representation of the density matrix between AO's μ ($\in$ the 0th cell) and ν ($\in$ the jth cell). The actual total density matrix $P_{\mu\nu}$ is obtained by summation of $P^j_{\mu\nu}$ from $j = 0$ to $\pm M$, where M denotes the range of the neighboring cell number considered to have an actual interaction with the 0th cell. Thus, the final set of CO's $\psi_s(k, x)$ and accompanied energies $\varepsilon_s(k)$ is obtained after the iterative process of larger scale compared with the usual MO calculation since the number of the Fock matrices to be solved is the same with the number of the k points considered in the Brillouin zone, which is usually 100 as mentioned above. Also, the total energy per unit cell E/N and the Mulliken populations, for instance, can be obtained from the final total density matrix element. In the optimization process of the 1D polymer, this E/N value is minimized with respect to all the atomic nuclei in the unit cell and the translation length. In the calculation method by the DFT framework, the basic scheme to obtain the CO's is essentially the same and the prescription of solving the Fock equations is changed into the KS equations mentioned in Sect. 3.2.

It is further noted that the CO $\psi_s(k, x)$ is a complex function, in general, having an imaginary part and, hence, that it is useless to draw their patterns unlike those of MO's being real function as seen, e.g., in Sect. 2.2. However, the highest occupied CO (HOCO) and the lowest unoccupied CO (LUCO) appear mostly at $k = 0$ or $\pm\frac{\pi}{a}$, and any CO's become real functions there, one can check the HOCO and the LUCO patterns in most cases. In Fig. 3.11, those patterns of polyacetylene and polythiophene are shown, for instance.

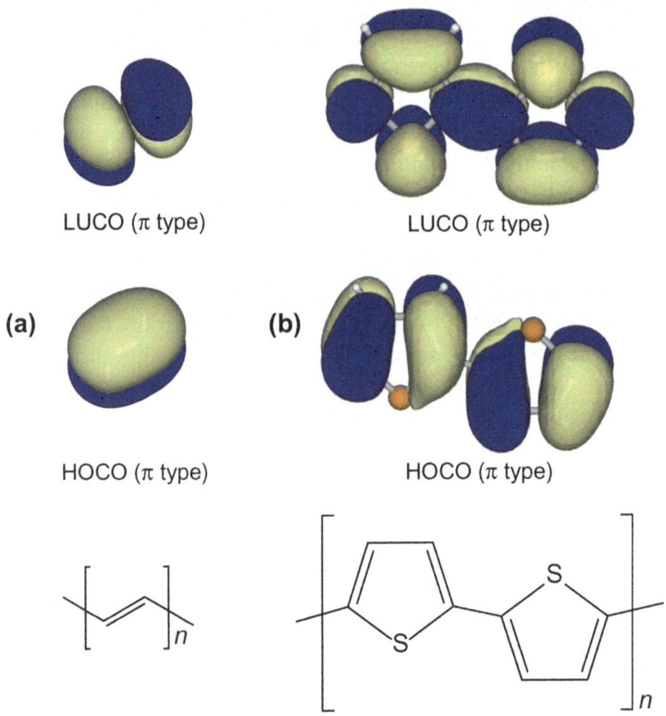

Fig. 3.11 The HOCO and the LUCO patterns of **a** polyacetylene and **b** polythiophene. They are drawn only in the unit cells. All of these are of π types

3.3.4 Energy Band

The CO expressed by Eq. (3.39) is accompanied by energies $\varepsilon_s(k)$ being a function of the wave vector k. One may sometimes see bizarre drawings named "energy band" or "band structure" for most of the crystal systems including a 1D polymer. This image is one of the typical and crucial key parts describing and characterizing the electronic properties of the 1D polymer as well. The energy band is a conventional expression method for the collection of energies $\varepsilon_s(k)$ with respect to each s and function of k in the Brillouin zone above mentioned.

A brief explanation for appearance of the energy band is as follows: in Fig. 3.12, the change of appearance of the MO energies is illustrated when a molecule "grows" into the 1D polymer. That is, when the repeated alignment of a unit cell along one direction reaches very long, the accompanied MO energies become densely distributed and finally continuous in the energy space. This actually corresponds to the origin of an energy band when one considers the infinite repetition of a simple unit cell. Hence, it is understood that the dense collection of MO energies is expanded to appear as a "band" along the wave vector k as seen in Fig. 3.13. The wave vector k also has a role of a new quantum number to specify the CO and its energy which constructs the energy band indexed by s in addition to representation of the momentum of an electron in the CO $\psi_s(k, x)$.

One can thus plot $\varepsilon_s(k)$ for each s, which eventually makes a collection of continuous curves called energy bands as illustrated in Fig. 3.14, where the valence bands and the conduction bands are indicated by those with $s = 1, 2$ and $s = 3, 4$, respectively. Among these, the band with $s = 2$ is called the highest occupied (HO) band and the one with $s = 3$ the lowest unoccupied (LU) band as also indicated in Fig. 3.14.

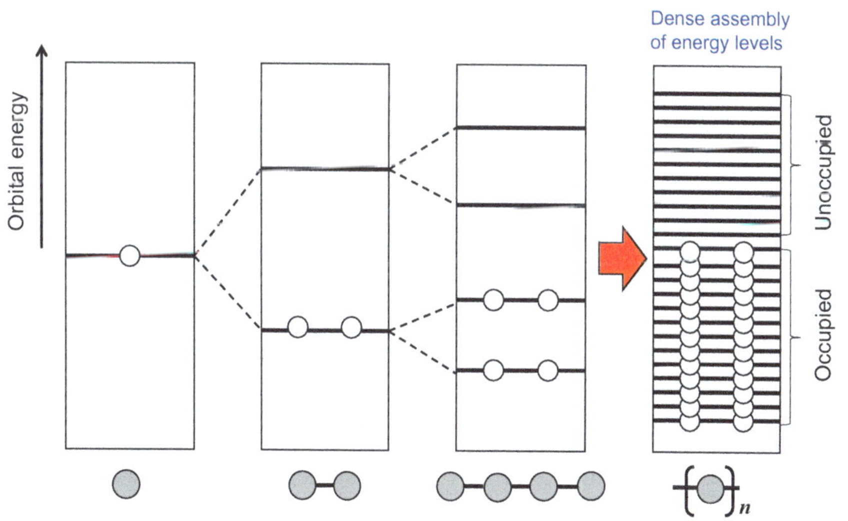

Fig. 3.12 Change of the orbital energy levels from an atom to a molecule and finally to 1D polymer

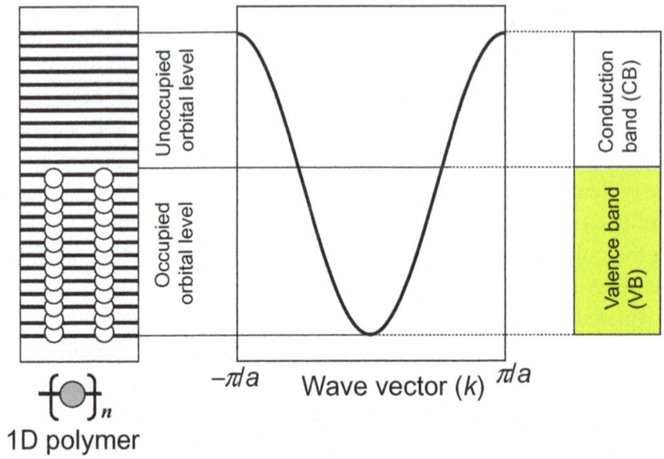

Fig. 3.13 Relationship of energy levels and energy band

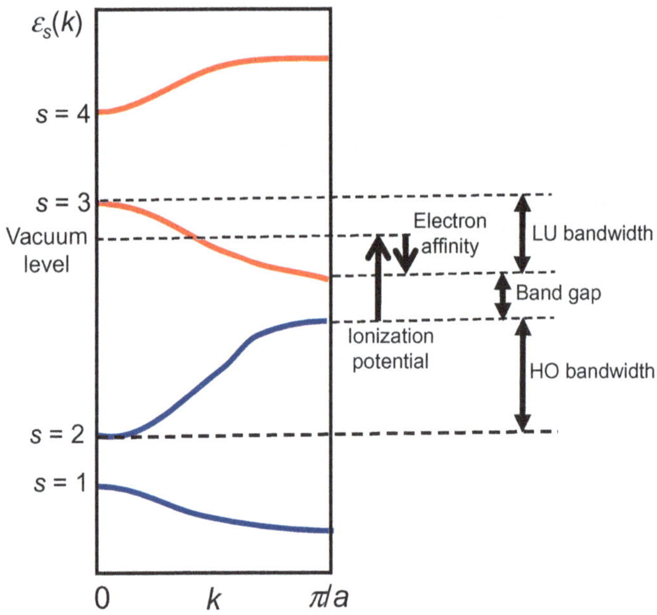

Fig. 3.14 Simplified picture of an energy band with the derived electronic properties

The top point of the HO band implies the energy of the HOCO and the bottom point of the LU band that of the LUCO. The energy bands are often plotted in the "half Brillouin zone", $\left[0 \leq k \leq \frac{\pi}{a}\right]$, since they have the mirror symmetry with respect to the plane at $k = 0$ due to energy degeneracies originating from the time-reversal

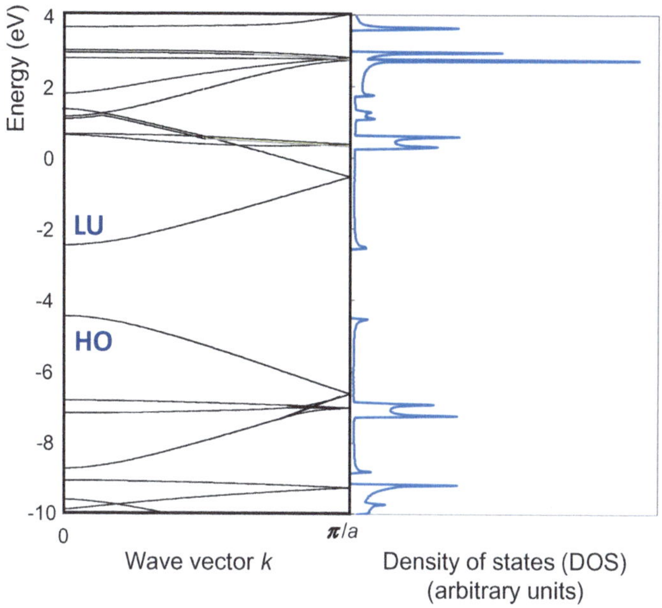

Fig. 3.15 Band structures and the density of states (DOS) of polythiophene in Fig. 2.7b. Note the unit of energy is expressed in eV here

symmetry. That is, an electron with momentum k has the same energy as that with $-k$ and, hence,

$$\varepsilon_s(k) = \varepsilon_s(-k) \tag{3.41}$$

for any band branches allotted by s. In this sense, it is sufficient to perform the ordinary calculation procedure within the half Brillouin zone. An example of the energy band structure (band structure) of polythiophene is shown in Fig. 3.15. It is noted that the gradient of each energy band becomes zero at the zone boundaries, i.e., $k = 0$ and $\pm\frac{\pi}{a}$ (Heine 1960).

The band structure affords miscellaneous information of several electronic properties of the 1D polymer as added in Fig. 3.14 similarly to those of molecules described in Sect. 2.4 as in what follows:

(1) Bandgap: Energy difference between the top of the HO band ($\varepsilon_{\mathrm{HO}}$) and the bottom of the LU band ($\varepsilon_{\mathrm{LU}}$) irrespective of k is called bandgap ΔE_{g}. This value gives the minimum excitation energy of the 1D polymer both thermally and optically. It is noted that the minimum optical excitation energy in the 1D polymer is also given simply by ΔE_{g} due to the vanishment of the electron repulsion integrals among the CO's. This is typically different from the excitation energies of molecules mentioned in Sect. 3.1.1. There can be direct and indirect bandgaps depending on the positional relationship of the band structures as seen

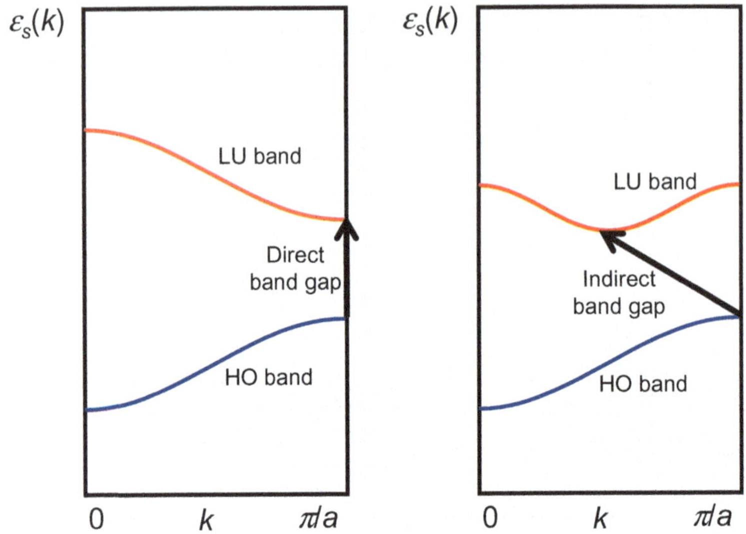

Fig. 3.16 Two kinds of bandgaps: **a** direct and **b** indirect

in Fig. 3.16. The bandgap in Fig. 3.14 happens to be direct one at the Brillouin zone boundary ($k = \pm\frac{\pi}{a}$). An indirect gap is based on the change in electron momentum $\hbar k$ so that it is associated with a forbidden transition. The bandgap is classified by the natures of the HO and LU bands and hence often called, for instance, the bandgap of $\sigma \rightarrow \sigma^*$, $\pi \rightarrow \pi^*$, or $n \rightarrow \pi^*$ type. The band-gap value can experimentally be obtained by measurement of the UV-vis absorption or by the redox potential.

(2) Ionization potential: Energy value of the top of the HO band with the opposite sign ($-\varepsilon_{\mathrm{HO}}$) is called ionization potential (I_{p}) as in the case of a molecule. This value affords information on the easiness of oxidation of the polymer. This value is experimentally measured by the ultraviolet photoelectron spectroscopy (UPS) or the electrochemical oxidation potential.

(3) Electron affinity: Energy value of the bottom of the LU band with the opposite sign ($-\varepsilon_{\mathrm{LU}}$) is called electron affinity (E_{A}) as in the case of a molecule. This value affords information on the easiness of reduction of the polymer. This value is hard to be spectroscopically measured but is electrochemically measured or with the combination of ΔE_{g} and I_{p}.

(4) Bandwidth: An energy range of each band is called the bandwidth. Often discussed are the HO and the LU bandwidths. The width value is associated with the degree of the spatial extension of the corresponding CO. That is,

large delocalization of electrons accommodated in the CO has a tendency of increase of the corresponding bandwidth. For instance, a large bandwidth of the band with π nature signifies the extension of the π conjugation throughout the polymer chain. This kind of extension is appropriate, e.g., for electrically conductive polymers. On the other hand, CO's accommodating σ electrons or lone-pair electrons are associated with the bands having relatively small bandwidth, and these energy bands are almost flat due to the spatial localization of those electrons.

(5) Effective mass of electron: This quantity implies the mass of electron or hole (carrier) running through the CO. This value may be smaller or larger compared with the actual mass of electron m_0 in vacuum depending on the characteristics of the CO's in the polymer. Effective mass of electron m^* is defined by

$$\frac{1}{m^*} = \frac{1}{\hbar^2} \frac{d^2 \varepsilon(k)}{dk^2} \tag{3.42}$$

Analysis of m^* becomes necessary to the discussion of the transport property of electrons in 1D polymer and hence m^* at the top of the HO band or the bottom of the LU band is often theoretically enumerated. The smaller absolute value of m^* signifies more mobile carriers. The negative value of m^* is often obtained at the top of the HO band, which can be interpreted as an effective mass of hole.

(6) Density of states (DOS): It is possible to define the density of states (DOS) by

$$N(\varepsilon) = C \frac{1}{\left(\frac{d\varepsilon(k)}{dk}\right)} \tag{3.43}$$

which represents the relative number of electrons at each energy value. That is, the DOS represents the distribution of $\varepsilon_s(k)$ and, in other words, the more number of $\varepsilon_s(k)$ in the energy interval, the larger becomes DOS. The constant C in the right-hand side is a sort of the normalization constant to regulate the number of total electrons per cm^3, g, and so on. The DOS of polythiophene is also shown in Fig. 3.15 adjacent to the band structure. It is found that the DOS steeply increases at the small gradient portion of the energy band due to the inversed differential in Eq. (3.43).

The actual example values of the electronic properties mentioned above are listed in Table 3.2.

It is noted that for the open-shell system as to the unit cell, the unrestricted method signifying different CO's for different spins is normally employed to give the split band structures for α and β spin electrons. Such a system is related to, e.g., the ferromagnetic polymers.

Table 3.2 Electronic properties of polyacetylene (*trans*-type) and polythiophene[1]

	Polyacetylene	Polythiophene
Bandgap (eV)	1.192	1.965
Ionization potential (eV)	4.032	4.486
Electron affinity (eV)	2.840	2.521
HO bandwidth (eV)	5.612	4.274
LU bandwidth (eV)	5.901	3.845
Effective mass at the top of the HO band[2]	$-0.047m_0$	$-0.144m_0$
Effective mass at the bottom of the LU band[2]	$0.046m_0$	$0.157m_0$

[1]The 1D polymer calculations including the structural optimization (see Fig. 2.7) were performed by DFT/B3LYP/6-31G**
[2]m_0 signifies the mass of a free electron in vacuum

3.4 Molecular Simulations

3.4.1 Outlook of Molecular Simulations

Molecular simulation is a kind of different area from quantum chemistry in that it deals with mostly the nuclear motion of molecules, oligomers, or even macromolecules including biological substances. The category of molecular simulation is usually divided into a couple of regions: molecular mechanics (MM), molecular dynamics (MD), and Monte Carlo (MC) methods. All of these do not explicitly examine the electronic structures of molecules but rather concentrate on nuclear motions usually under certain empirical potential energy functions.

The above three methods have their own features and characteristics. Among them, the MM method is rather simple and convenient without invoking a huge amount of calculations including electrons like quantum chemical ones, so that it can easily provide conformational analyses of molecular structure. The MD method mostly traces the motion of particles by solving the classical Newtonian equations of motion of atoms or molecules with minute time intervals for extended periods of time. This method affords information of “trajectory” of the whole system so as to take the time average. Finally, the MC method employs the statistical mechanics procedure for the description of the transition of atoms or molecules in order to acquire large numbers of “configurations” giving their positions like snapshots in order to take the ensemble average without direct solving the equations of motion. In this section, the MM, MD, and MC methods are briefly introduced.

3.4.2 Molecular Mechanics (MM)

The MM method consists of a rather simple idea in that conformation of molecules, oligomers, or even longer polymeric chains is determined by the molecular motion under the empirical potential energy function employed (Kollman et al. 2000). Part of this method has already been mentioned in Sect. 2.1 used in the preliminary optimization process of molecular structure. The most important ingredient for MM calculation is the selection of appropriate empirical potential energy function $V(\mathbf{R}_1, \mathbf{R}_2, \ldots, \mathbf{R}_N)$ applying to the molecule or polymer chains consisting of N nuclei, where $\mathbf{R}_i$ implies the vector coordinate of the ith nucleus. The potential energy V can be expressed by the summation of several contribution terms such as follows:

$$V = V_1 + V_2 + V_3 + V_4 + V_5 \tag{3.44}$$

each V_i being described as follows:

$$V_1 = \sum_{\text{bond}} \frac{1}{2} k_r (r - r_0)^2 \quad \text{(bond-length stretching)} \tag{3.45}$$

$$V_2 = \sum_{\text{angle}} \frac{1}{2} k_\theta (\theta - \theta_0)^2 \quad \text{(bond-angle deformation)} \tag{3.46}$$

$$V_3 = \sum_{\text{dihedral angle}} \frac{1}{2} k_\varphi \{1 + \cos(n\varphi - \gamma)\} \quad \text{(dihedral-angle deformation)} \tag{3.47}$$

$$V_4 = \sum_{\text{angle}} \frac{1}{2} k_\rho \rho^2 \quad \text{(out-of-plane angle deformation)} \tag{3.48}$$

$$V_5 = \sum_{i>j} \left[\frac{A_{ij}}{r_{ij}^{12}} - \frac{B_{ij}}{r_{ij}^{6}} + \frac{q_i q_j}{\varepsilon r_{ij}} \right] \quad \text{(long-range interaction)} \tag{3.49}$$

where each variable is illustrated in Fig. 3.17. In these equations, $k_r \sim k_\rho$ are the force constants decided in the individual MM method, and the subscript 0 to each variable expresses the value at the equilibrium position. The values n and γ in Eq. (3.47) signify, respectively, the numbers of local minima and phase angle taking the value of either 0° or 180°. In Eq. (3.49), A_{ij} and B_{ij} are the constants for van der Waals and London dispersion interactions, respectively, and q_i and ε being the ith partial atomic charge and dielectric constant of the surrounding medium.

The process of the MM calculation is to start from setting the initial coordinates of n molecules $(\mathbf{R}_1, \mathbf{R}_2, \ldots, \mathbf{R}_N)$ and then calculate the potential energy $V(\mathbf{R}_1, \mathbf{R}_2, \ldots, \mathbf{R}_N)$ felt by those molecules. The following procedure is quite similar to the structural optimization of molecules described in Sect. 2.1. The only difference

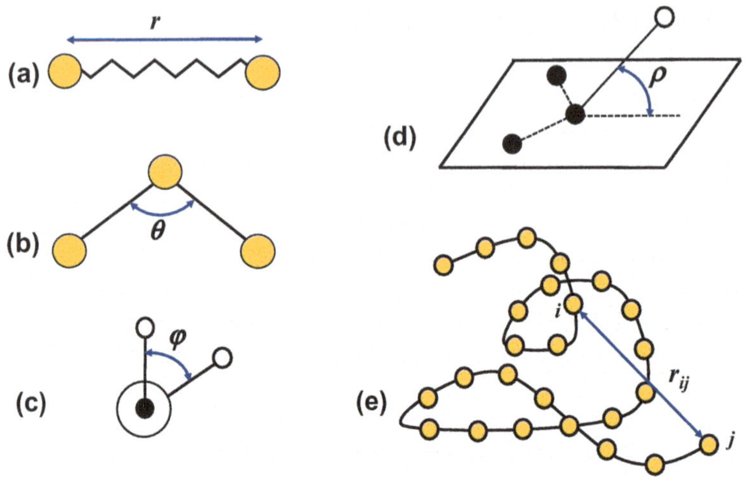

Fig. 3.17 Variables for energy potential. **a** Bond length r, **b** bond angle θ, **c** dihedral angle φ, **d** out-of-plane angle ρ, and **e** long-range interaction r_{ij}

is to employ an economical and tractable potential energy function in the MM calculation. Thus, the energy obtained in the MM scheme is not the total energy including the kinetic term but only the potential energy which is usually called "steric energy". It is noted that the normal vibration analysis can be performed since the force constants are obtained from the Hessian matrix and acquisition of the thermochemical data such as enthalpy, entropy, and Gibbs free energy becomes available as well. Hence, this thermochemical information affords possibility toward the conformational search of molecules having several isomeric conformers.

Targets of the MM calculation normally include molecules, oligomers, and especially those with rather long chains and hence, also applicable to biomolecules. There have been developed several force fields or, in other words, types of potential energy functions employed in the MM calculations called, e.g., MM2, MM3, MMFF94, AMBER, CHARMM, and so on with rather individual potential energy function. Parametrizations in empirical potential energy function are uniquely decided in each software.

There have been published several MD softwares as in what follows with their URL's (which may change).

Free softwares:

GROMACS (Groningen Machine for Chemical Simulations) http://www.gromacs.org/
OCTA http://octa.jp/

Paywares:

Chem3D http://www.cambridgesoft.com/

AMBER (Assisted Model Building with Energy Refinement) http://ambermd.org/
CHARMM (Chemistry at Harvard Macromolecular Mechanics) Partially free. https://www.charmm.org/charmm/
CONFLEX http://www.conflex.net/

3.4.3 Molecular Dynamics (MD)

In the MD simulation of a system consisting of N particles, the equations of motion are explicitly considered in the framework of classical mechanics described as follows:

$$M_i \frac{\mathrm{d}^2\mathbf{R}_i(t)}{\mathrm{d}t^2} = \mathbf{F}_i(t) \quad (i = 1, 2, 3, \ldots, N) \tag{3.50}$$

where M_i, $\mathbf{R}_i$, and $\mathbf{F}_i$ stand for, respectively, the mass, position vector, and force applied concerning the ith particle. These equations are numerically solved based on the difference method, in which, starting from the initial condition (information on $\mathbf{R}_i(0)$ and $\mathbf{v}_i(0)$, where $\mathbf{v}_i(t)$ stands for the velocity of the ith particle), each equation is solved stepwise for every time interval Δt as illustrated in Fig. 3.18. In the difference method, $\mathbf{R}_i(t + \Delta t)$ is expanded into the Taylor series usually up to the second order to give

$$\mathbf{R}_i(t + \Delta t) - \mathbf{R}_i(t) = \frac{\mathrm{d}\mathbf{R}_i(t)}{\mathrm{d}t}\Delta t + \frac{1}{2}\frac{\mathrm{d}^2\mathbf{R}_i(t)}{\mathrm{d}t^2}\Delta t^2 = \mathbf{v}_i(t)\Delta t + \frac{\mathbf{F}_i(t)}{2M_i}\Delta t^2 \tag{3.51}$$

The time interval Δt is taken to be shorter than the motion of a particle, i.e., ca. 0.1 fs (1 femtosec = 10^{-15} s). The force $\mathbf{F}_i$(t) is usually taken to be time-independent and is duly obtained by the expression

$$\mathbf{F}_i(t) = \mathbf{F}_i = -\nabla_i V(\mathbf{R}_1, \mathbf{R}_2, \ldots, \mathbf{R}_N) = \left(-\frac{\partial V}{\partial x_i}, -\frac{\partial V}{\partial y_i}, -\frac{\partial V}{\partial z_i}\right) \tag{3.52}$$

where $V(\mathbf{R}_1, \mathbf{R}_2, \ldots, \mathbf{R}_N)$ stands for the empirical potential energy function for N particles or is called a molecular force field similar to what is used in the MM method.

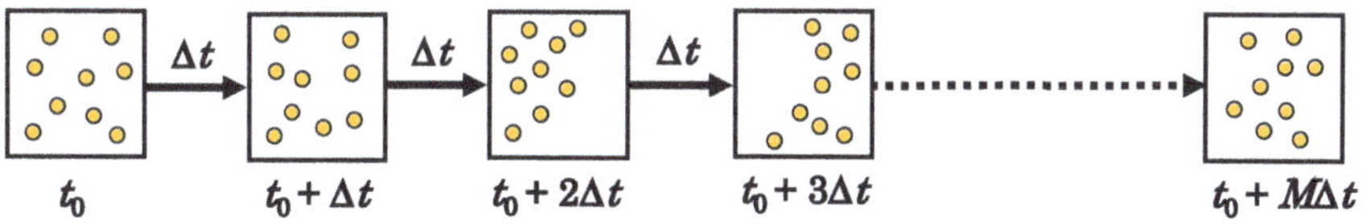

Fig. 3.18 Concept of the MD method, which eventually generates M samples of the molecular state (M $\gg$ 1) along with the time evolution with each time step

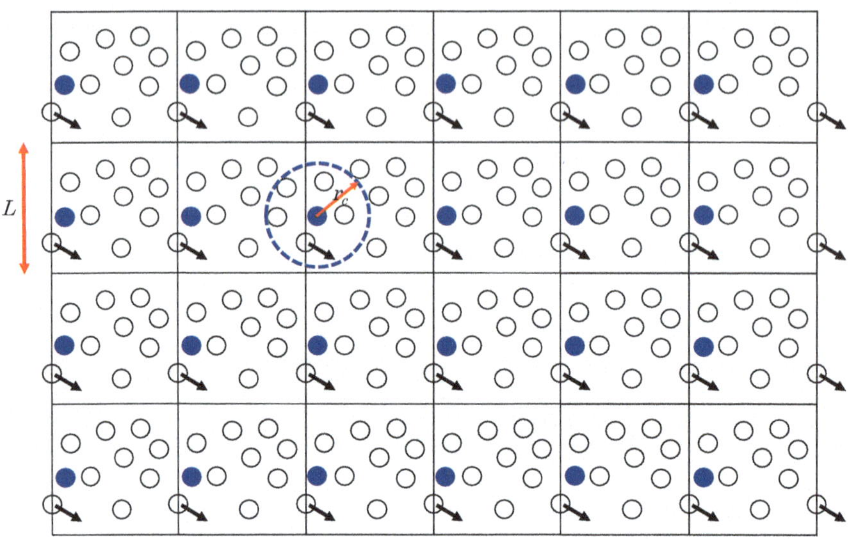

Fig. 3.19 Image of a periodic boundary condition in the MD method. Each square represents the unit cell with each side L. The broken circle indicates the potential cut off range r_c

In recent years, there are also possibilities for employing quantum mechanical estimation for electronic structures included in $V(\mathbf{R}_1, \mathbf{R}_2, \ldots, \mathbf{R}_N)$ which is called quantum mechanical/molecular mechanical (QM/MM) approximation for MD method.

Since the MD method deals with the equations of motion of particles, time-dependent behaviors of those can be traced, from which the term "molecular dynamics" has resulted (Alder and Wainwright 1959; Frenkel and Smit 2001). Moreover, in the MD calculation, it is usual to increase the number of particles dealt with up to virtually infinite by employing the periodic boundary condition as illustrated in Fig. 3.19, in which the repeating unit cell usually involves 10^4–10^5 particles depending on the capacity of the computer used. This condition is somewhat similar to that utilized in Sect. 3.3 for the treatment of electronic structures of polymers but differs in that the particles are incessantly moving inside the unit cell. Moreover, when a particle hit against wall of the unit cell it shall instantly appear from the wall of the neighboring unit cell as an illustration of the arrows in Fig. 3.19 implies. This concept guarantees the conservation of the total momenta of particles in the unit cell. Furthermore, based on the consideration of the interaction between the particles inside and outside the unit cell, those existing in the potential cut off range r_c ($r_c < L/2$; L being each side of a unit cell) as shown in Fig. 3.19 shall be counted to eventually conserve the total energy in the unit cell.

Note that we reserve the particle number N, the total energy E, and, implicitly, the volume V of the system as described hitherto. The MD analysis ranging over the long-time period will eventually bring the system to the equilibrium state regardless of the initial condition by $\mathbf{R}_i(0)$ and $\mathbf{v}_i(0)$, which enables to stabilize physical quantity A such as the temperature T, pressure P, or other thermodynamical variables in the sense of their time average

$$\langle A \rangle_t = \lim_{t \to \infty} \frac{1}{t} \int_0^t A(t)\mathrm{d}t \simeq \frac{1}{M} \sum_{n=1}^{M} A(t_0 + n\Delta t) \tag{3.53}$$

where M signifies the sample numbers which should be sufficiently large along the time evolution, at each point of which the MD equation is solved. One can also set a specific ensemble of particles with constant combinations of (N, V, E), (N, V, T), or others in the MD analysis by retaining the corresponding external conditions. MD analyses are often applied to the molecular motion connecting to a chemical reaction in solution, discussion of active sites and function of biopolymers, and even to a non-equilibrium condition which ensures estimation of energy propagation and relaxation mechanism of heat. Some specific examples of the MD calculations will be given in Sect. 4.4.

There have been published several MD softwares.

Free softwares:

GROMACS (Groningen Machine for Chemical Simulations) http://www.gromacs.org/
OCTA http://octa.jp/

Paywares:

Chem3D http://www.cambridgesoft.com/
AMBER (Assisted Model Building with Energy Refinement) http://ambermd.org/
CHARMM (Chemistry at Harvard Macromolecular Mechanics) Partially free. https://www.charmm.org/charmm/
HyperChem http://www.hyper.com/

3.4.4 *Monte Carlo (MC) Method*

The basic idea of MC method lies in the statistical-mechanics procedure and does not directly deal with the equation of motion of molecules nor trace the time evolution as is performed in the MD analysis. In other words, molecules dealt with in the MC calculation are considered to be located under the potential energy as least as possible without showing an image of their velocity nor momentum. In this sense, MC analysis is not appropriate to discuss events with the time evolution but is apt to describe the stationary ensemble characterization.

Typical prescriptions leading to the equilibrium state of an (N, V, T) ensemble consisting of molecules is as follows (Metropolis et al. 1953):

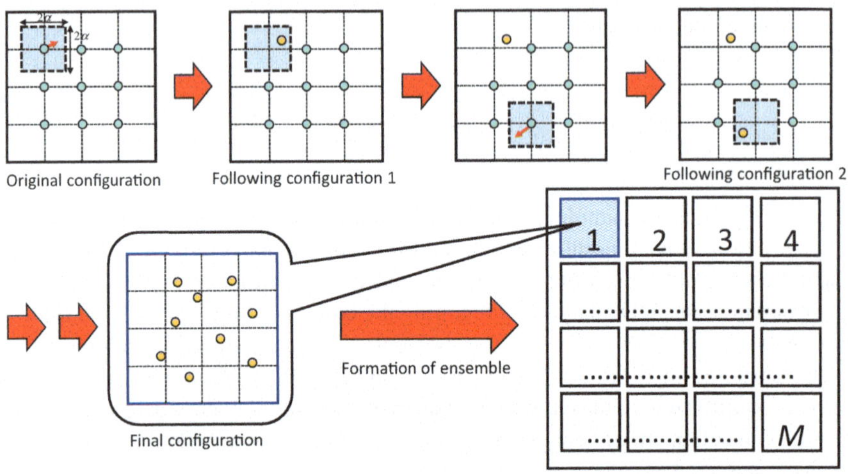

Fig. 3.20 Block diagram of Metropolis method to eventually generate an (N, V, T) ensemble consisting of M configurations ($M \gg 1$) for the MC calculation. See the text as to 2α in the original configuration

(1) Set molecule numbers N (10^5–10^6) involved in the concerning system and the temperature T therein.
(2) Consider a unit cell of volume V on which the periodic boundary condition is imposed as in the MD method. In this unit cell, N molecules shall be located arbitrarily as the first configuration, the total potential energy of which is calculated using the empirical potential energy functions as in the MM method. In this calculation, only the long-range interaction term in Eq. (3.49) as the potential energy is crucial since the bond-length stretchings and bond-angle deformations are usually not considered in the MC process.
(3) A molecule in the above unit cell is selected to randomly change its position within a definite size of, say, $2\alpha \times 2\alpha$ square with the aid of random numbers as in Fig. 3.20 centered at the original point where this molecule is. In this procedure, not only changing the position of the molecule but also the molecular rotation can be considered at the same time.
(4) Potential energy is newly calculated as to the unit cell with an altered position of the molecule described in (3) as the "next configuration". When the potential energy is *stabilized* in this next configuration, it is accepted as the "following configuration".
(5) On the other hand, in case there is potential energy destabilization, say by ΔE (> 0), the above new configuration is accepted as the "following configuration" only when

$$\xi > \exp\left[-\frac{\Delta E}{k_B T}\right] \tag{3.54}$$

holds, where ξ is a pre-set random number satisfying the range $0 < \xi < 1$. On the other hand, when Eq. (3.54) is not satisfied, the original configuration is adopted as the following configuration.

(6) The processes (3)–(5) for other molecules in the unit cell is repeated, therefore, totally about 10^5–10^6 times, and the eventual configuration itself achieves the energetically stable state.

(7) Furthermore, ca. 10^6 (say M) of the energetically stable configurations are obtained to finally form an (N, V, T) ensemble in total, which is used to calculate the ensemble average for a physical quantity A simply expressed by

$$\langle A \rangle = \frac{1}{M} \sum_{j=1}^{M} A\left(\mathbf{R}_1^{(j)}, \mathbf{R}_2^{(j)}, \ldots, \mathbf{R}_N^{(j)}\right) \tag{3.55}$$

where the value A is determined by the positions of N molecules through the potential energy function $V(\mathbf{R}_1, \mathbf{R}_2, \ldots, \mathbf{R}_N)$ with j specifying each configuration in Fig. 3.20.

In a more formal description Eq. (3.55) can be rewritten as follows:

$$\langle A \rangle = \int A(\mathbf{R}_1, \mathbf{R}_2, \ldots, \mathbf{R}_N) f(\mathbf{R}_1, \mathbf{R}_2, \ldots, \mathbf{R}_N) \mathrm{d}\mathbf{R}^N \tag{3.56}$$

using the distribution function f often used in statistical mechanics. This f actually signifies the Boltzmann distribution given by

$$f(\mathbf{R}_1, \mathbf{R}_2, \ldots, \mathbf{R}_N) = \frac{\exp\left[-\frac{V(\mathbf{R}_1, \mathbf{R}_2, \ldots, \mathbf{R}_N)}{k_B T}\right]}{Z_N} \tag{3.57}$$

for N molecules, where Z_N is called partition function and acting as the normalization factor for the distribution f.

With the above Metropolis scheme, the obtained (N, V, T) ensemble would be considered to satisfy the ergodic hypothesis. Under this condition, the ensemble average of physical quantity A is expected to become equal to its time average defined in Eq. (3.53), i.e.,

$$\langle A \rangle = \langle A \rangle_t \tag{3.58}$$

The general reference for the MC method is elsewhere (Jorgensen and Tirado-Rives 2005). At present, not so many softwares for the MC method have been published and it seems each is rather house-made.

Free software:

BOSS (Biochemical and Organic Simulation System).
http://zarbi.chem.yale.edu/software.html.

3.5 ONIOM Method

3.5.1 Concept of the ONIOM

For large or huge molecules such as polypeptides or nucleic acids, the orthodox MO calculation would be too much time consuming and may not be practical. In such a case, it is sometimes convenient to partition the whole molecule into a couple of "layers" and choose the degree of precision of MO calculation for each layer.

One of such and popular frameworks is ONIOM (Our own n-layered integrated molecular orbital and molecular mechanics) method (Svensson et al. 1999), in which the molecule concerned is divided into mostly two or three layers as illustrated in Figs. 3.21 and 3.22, which is called ONIOM2 and ONIOM3, respectively. The first[t] layer is the most important portion of the molecule and is also called the Model system, whereas the whole molecule Real system in the ONIOM method.

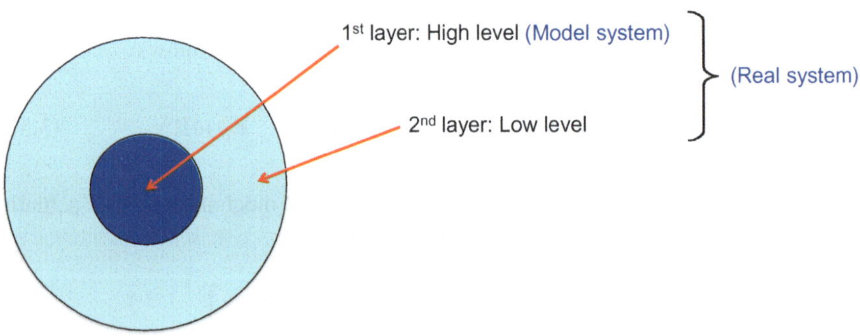

Fig. 3.21 Concept of the layers in the ONIOM2 method. High and low indicate the degrees of precision by employing different calculation levels. It is not necessary that these layers should be of sphere-like structures

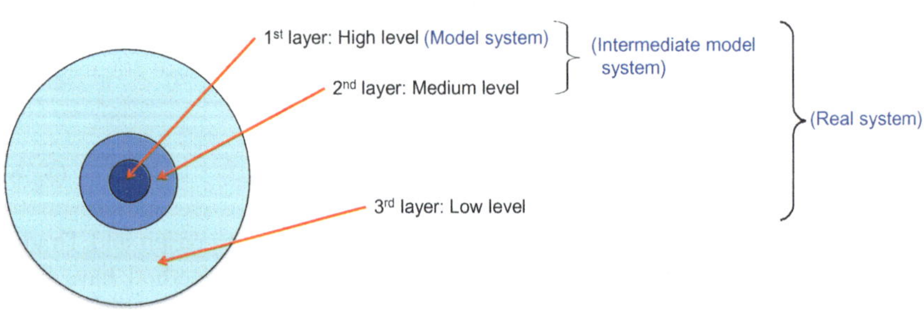

Fig. 3.22 Concept of the layers in the ONIOM3 method. High, medium, and low indicate the degrees of precision by employing different calculation levels. It is not necessary that these layers should be of sphere-like structures

For the 1st layer, high level of the MO calculation is applied, whereas the 2nd and the 3rd layers are less important in order so that intermediate- and low-level calculations are employed for these. These situations are schematically shown also in Figs. 3.21 and 3.22. Note that there is some difference in the calculation level for the 2nd layer depending on ONIOM2 or ONIOM3 method. For the 2nd and the 3rd layers, it is usual to employ the MO calculations with a simpler basis set or the semiempirical MO scheme, which is totally called QM/QM framework. Even the MM scheme might be applied for the low-level calculation called QM/MM framework.

For the two-layer ONIOM (ONIOM2) method, the energy of the real system is approximated by

$$E(\text{Real, High}) = E(\text{Real, Low}) + \{E(\text{Model, High}) - E(\text{Model, Low})\} \quad (3.59)$$

and for the three-layer ONIOM (ONIOM3) method by

$$\begin{aligned} E(\text{Real, High}) = {} & E(\text{Real, Low}) + \{E(\text{Intermediate, Medium}) - E(\text{Intermediate, Low})\} \\ & + \{E(\text{Model, High}) - E(\text{Model, Medium})\} \end{aligned} \quad (3.60)$$

These approximations will be simply extended to multilayer multilevel structures in the ONIOM framework.

It is most crucial to appropriately deal with the interlayer connection between both the layers in a seamless manner. Careful manipulations concerning these have been described in the earlier references (Maseras and Morokuma 1995; Svensson et al. 1996). The total energy, the energy gradient, most of the general electronic properties, and the vibrational properties of the Real system can thus be well retrieved so that these properties are acquired in a similar fashion to the ordinary calculation of the whole molecule. In other words, molecular structural optimization and search of the transition state during the chemical reaction can be performed are also attainable in the ONIOM calculation.

3.5.2 *Simple Example of the ONIOM Method*

Let us examine here a simple ONIOM2 calculation taking 1-propanol molecule as an example for the Real system and setting methanol part as the 1st layer and ethyl part the 2nd layer as shown in Fig. 3.23. Several combinations of the calculation method for each layer have been employed, the results of which are listed in Table 3.3. As the benchmark test CCSD/6-311G** is employed for both the 1st and the 2nd layers. Other calculation levels common to both layers are also attempted. It is natural that the CCSD/6-311G** calculation should take the longest computation time among others but gives the most stable energy implying the relatively precise result among the present calculations as expected. As the ONIOM2 calculation combination of the CCSD/6-311G**: MP2/6-31G* or the CCSD/6-311G**: HF/STO-3G gives a

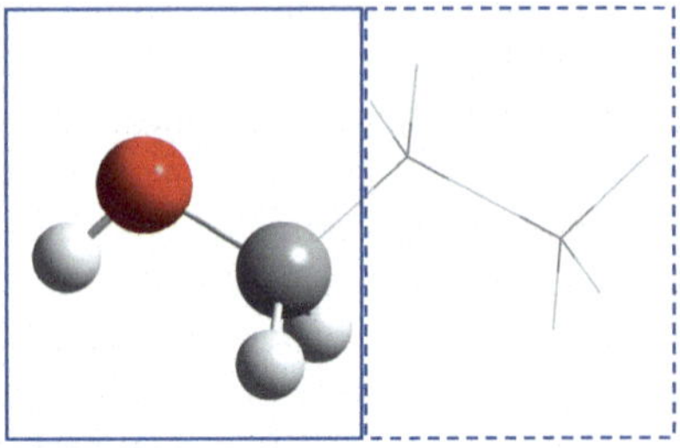

Fig. 3.23 A simple example of the application of ONIOM method to 1-propanol (*trans*-conformation). The solid rectangle indicates the first layer and dashed one the second layer

Table 3.3 Comparison of the ONIOM2 results for 1-propanol

Combination in the ONIOM2 method[1] (1st layer: 2nd layer)	Energy (in au)[2]	Dipole moment (in Debye)	Computation time
Benchmark CCSD/6-311G**: CCSD/6-311G**	−193.877793	1.535941	15 h 19 min
MP2/6-31G*: MP2/6-31G*	−193.682294	1.721306	1 min 10 s
HF/6-31G**: HF/6-31G**	−193.127959	1.620955	21.0 s
HF/6-31G*: HF/6-31G*	−193.110505	1.649561	10.0 s
HF/STO-3G: HF/STO-3G	−190.712911	1.405597	2.0 s
CCSD/6-311G**: MP2/6-31G*	−193.794593	1.516401	35 min 37 s
CCSD/6-311G**: HF/STO-3G	−192.622496	1.646425	27 min 15 s
HF/6-31G**: PM6 [3]	−115.064072 [5]	1.675196	7.0 s
HF/6-31G**: UFF [4]	−115.044284 [5]	1.851395	4.0 s

[1]In all of the calculations the initial molecular structure is pre-optimized by the MM2 method providing the *trans* (*anti*) conformation of 1-propanol
[2]At the optimized structures for each ONIOM combination
[3]Parametrized Model number 6 calculation in the category of semiempirical MO method
[4]Universal Force Field calculation in the category of the MM method
[5]Energy is incomparable with those in the above lines due to a lack of contribution from the inner electrons in the second layer calculation

reasonably good result with relatively shorter time compared with the benchmark test as seen in Table 3.3. Utilization of the semiempirical MO or MM calculation for the lower level requires far less computation time, which ensures effectiveness in large molecules. The latter calculation corresponds to, what is called, the QM/MM

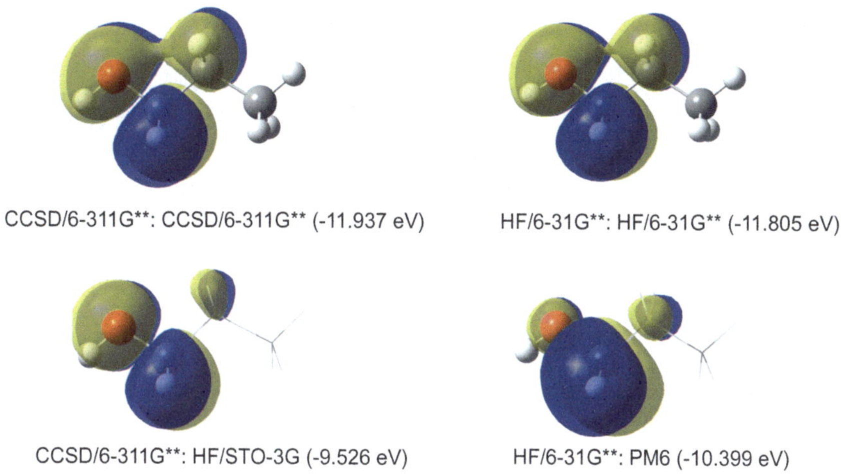

Fig. 3.24 Comparison of the HOMO patterns and their energy levels (in parentheses) of *trans*-1-propanol obtained by the ONIOM2 method with several combinations of the calculation. The moiety for the second layer calculation is represented by wireframe

framework, which makes it possible to examine the relative stabilization energies among the various conformations of a large Real system (see below).

The electronic properties such as, for instance, dipole moment and the MO patterns, in addition to the molecular energy are also available. Values of the dipole moment obtained by several ONIOM combinations are also listed in Table 3.3 is in reasonable agreement with the experimental value 1.55 Debye (Rumble et al. 2018). The HOMO patterns also obtained by these ONIOM combinations in Fig. 3.24 look similar but are a bit separated between the first and the second layers when different calculation levels are adopted for those. In the QM/MM scheme, the MO patterns are unavailable since the MM does not deal with the wavefunctions. It is further noted that, although the detailed data are omitted here, the IR frequencies can also be obtained since the ONIOM method is capable to afford the vibration analysis both in the QM/QM and QM/MM frameworks.

3.5.3 Further Examples

Below, a couple of more examples along with the original purpose of the ONIOM method toward application to a larger size of molecules or molecular systems are to be described. For instance, the QM/MM ONIOM2 method (B3LYP/SDD(Fe), 6-31G*(rest): AMBER) has been employed to a metalloprotein, soybean lipoxygenase-1 (SLO-1) with 839 amino acid residues in the quintet ground state resulting in that two conformers A and B each optimized (Fig. 3.25) have turned out to have similar energetical stabilities (within 1 kcal/mol) (Hirao and Morokuma 2010). It has also

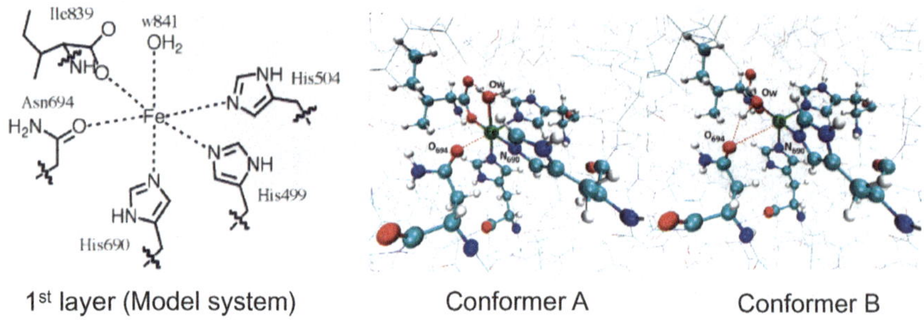

Fig. 3.25 The employed first layer of soybean lipoxygenase-1 (SLO-1) in the present ONIOM2 method with the boundaries of the first layer indicated by wavy lines. The two conformers A and B including the first layer have been obtained by energetical optimization. Reprinted with permission from Hirao and Morokuma (2010). Copyright 2010 American Chemical Society

been concluded that the average of the Fe–O_{694} distances in conformer A (2.39 Å) and B (3.46 Å) well agrees with that of the crystal structure. As a whole, an averaged geometry of A and B was in good agreement with the experimental result.

Microsolvated effects on the excitation energies of 2-aminopurine surrounded by water molecules have been examined by using QM/QM ONIOM2 (TD-B3LYP/6-31++G*: PM3). In both of the ground and the first-excited states, six water molecules interact with 2-aminopurine by hydrogen bonding as shown in Fig. 3.26 (Zhang et al. 2004). The first and the second lowest excitation energies obtained (4.29 and 4.43 eV assigned for $\pi \rightarrow \pi^*$ and $n \rightarrow \pi^*$, respectively) in aqueous medium well agree with the experimental results (4.11 and 4.46 eV). Moreover, this ONIOM method has predicted fluorescent emission energy for the $S_1 \rightarrow S_0$ de-excitation as 3.87 eV compared with the experimental value 3.70 eV based on the least square fitting for the experimental transition.

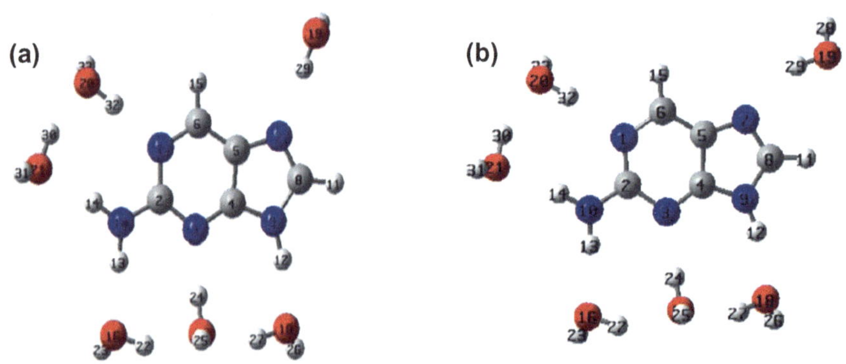

Fig. 3.26 Geometries of 2-aminopurine surrounded by six water molecules in **a** the ground state and **b** the lowest excited state obtained by the ONIOM2 method. The first layer contains a 2-aminopurine and the second layer six water molecules. Reprinted from Zheng et al. (2004). Copyright 2004 with permission from Elsevier

Physisorption of a water molecule on the modeled graphite ($C_{96}H_{24}$) has been examined by QM/QM ONIOM2 (B3LYP/6-31 + G*: DFTB-D) finding weak binding energy 1.8 kcal/mol, where DFTB-D signifies dispersion-augmented density functional tight-binding calculation (Xu et al. 2005). When five water molecules make a cluster the binding energy becomes 41.7 kcal/mol (Fig. 3.27) compared with the unbound and unclustered state, signifying there are obviously generated hydrogen bondings among the water molecules.

It is noted that the ONIOM3 method with a gradual change in the degrees of precision of the calculation for each layer might be also useful. In Fig. 3.28, the picture of solvation cluster for Li^+ with propylene carbonate (PC) as the solvent by the QM/QM/MM ONIOM3 calculation is shown. As the calculation scheme for each layer is employed, the B3LYP/6-31G: HF/STO-3G: UFF combination. The 1st layer calculation is performed for only a Li^+, the 2nd layer for the directly solvated four PC molecules (see Sect. 4.2; Yanase and Oi 2002 and Ohtani et al. 2010), and the 3rd layer for surrounding 173 PC molecules. The ONIOM3-optimized solvation cluster constructs approximately a sphere with the radius of ca. 21.5 Å, signifying the concentration of electrolyte solution of 0.040 M. Some data concerning this solvation model is listed in Table 3.4 in which the net charge of Li^+, averaged net charges of four oxygen atoms near to the central Li^+, and Li^+–O averaged bond orders do not remarkably change with the addition of the 3rd layer consisting of more than 170 PC molecules. On the other hand, Li^+–O bond lengths considerably change according to the enlargement of the PC solvation cluster. This signifies that the direct solvation cluster described by the 2nd layer tends to shrink when farther solvent molecules constructing the 3rd layer are taken into consideration.

Finally, it has generally been pointed out in the ONIOM method that sufficient preliminary examination is generally recommended to reduce possible mismatching

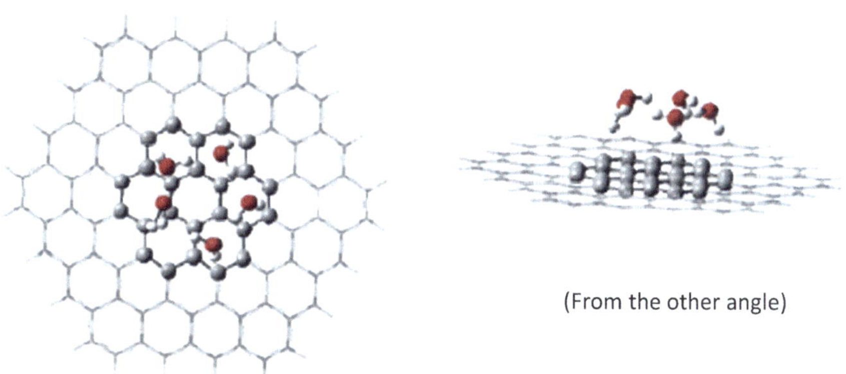

Fig. 3.27 The ONIOM2 calculation result of a supermolecule structure consisting of a modeled graphite ($C_{96}H_{24}$) and physisorbed water-molecule cluster ($(H_2O)_n$; $n = 5$) above the basal plane. Atoms expressed by spheres construct the first layer, and wireframe represents the second layer. Reprinted with permission from Xu et at. (2005). Copyright 2005 American Chemical Society

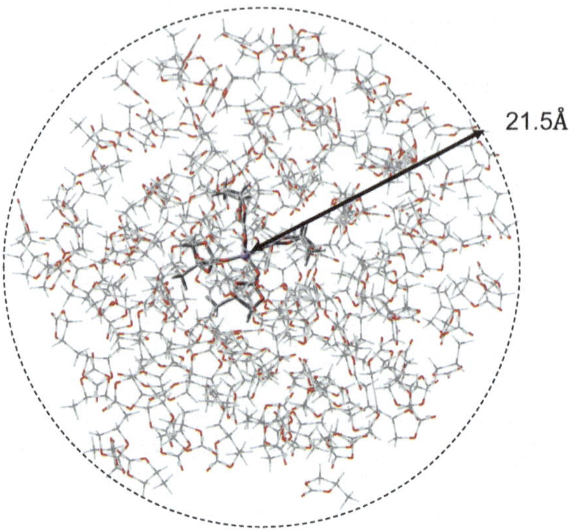

Fig. 3.28 The ONIOM3 calculation result of a solvation structure around Li^+ with propylene carbonate (PC) molecules. The central Li^+ (shown by purple) is taken as the first layer, the directly coordinating four PC molecules (shown by thick solid lines) as the second layer, and the farther 173 PC molecules indicated by wireframe as the third layer. The total solvation cluster forms a sphere, the radius of which is ca. 21.5 Å

Table 3.4 The ONIOM3 results for Li^+ solvated with propylene carbonate (PC) molecules

	Li^+–O averaged bond lengths (in Å)	Li^+ net charge[1]	O averaged net charges[1]	Li^+–O averaged bond order[1]
Up to the 2nd layer: Li^+-PC_4	1.921	+0.4996	−0.4609	0.1185
Up to the 3rd layer: Li^+-PC_4-PC_{173}	1.783	+0.4283	−0.4693	0.1190

[1]Due to Mulliken population analysis

of chemical bonding and/or unbalanced combination of the calculation methods and the basis sets at the layer boundaries.

3.6 Hints for Calculations

There are various calculation methods employed in theoretical chemistry at present as have been explained in Sects. 3.1–3.5. This comes from a great deal of efforts and improvements that have been continuously accumulated and improved hitherto toward the development for satisfying higher precision, adaptation for larger molecule size, and faster processing capacity. Along with this process, a plenty of computation softwares have been published for the execution of calculations fulfilling those purposes. On the other hand, theoretical-calculation users might be more and more embarrassed or left in confusion toward the selection of appropriate method out of the current calculation resources for their own purpose to obtain the effective results to reveal, support, or augment the points they need to know in the course of the experimental research.

There can be miscellaneous chemical subjects for the users to search for as a matter of course. These demands would be related to, for instance, structure, electronic properties, chemical reactivities, and so on with respect to molecules and polymers. In this section, we attempt to afford certain hints and/or prescriptions using the Q & A style toward the effective selection of the calculation method from practical points of view.

Q1: What kind of theoretical calculations can be used for molecules?

A1: There are several categories of calculation methods as summarized below. Of these should be employed appropriate one (ones) considering motivation and demand in one's research, since there are certain characteristics and differences in these methods including reasonable cost performances such as computation time and memory size.

(1) Molecular orbital (MO) method

This is to analyze the electronic structure of molecules in terms of the MO. The MO can also be extended to 1D polymer with the introduction of special boundary conditions to the MO's to the CO's. A detailed calculation scheme in the MO or the CO method can be roughly classified into the following four:

Empirical scheme:
This scheme includes the Hückel and extended Hückel methods. Common features

of these are that the computation time is extremely short and the output is rather qualitative. The former method is able to deal with only π electrons and the latter both σ and π electrons. It is noted that these methods cannot discuss the energy of molecules due to their theoretically simple frameworks. However, the MO patterns obtained are sometimes useful for a qualitative discussion of chemical reaction of the concerning molecule(s).

HF scheme:

This can be classified into the semiempirical and the non-empirical (also used to be called ab initio) HF methods (see Sect. 3.1.1). The semiempirical method is equipped with several entries such as CNDO, INDO, MINDO, CNDO/S, ZINDO, MNDO, AM1, PM6, and others, all of which deal with only the valence electrons requiring quite short computation time and giving qualitative results. The semiempirical method is sometimes convenient for certain purposes to well reproduce the excitation energies, for instance, due to the employment of the parameters fitted to the experimental data.

On the other hand, the non-empirical method is characterized by that it does not contain any experimental parameters, while it requires rather longer computation time and gives semiquantitative result largely depending on the basis sets employed.

Post-HF scheme:

This is the improved version of the HF method with the inclusion of an electron-correlation effect and is expected to afford more quantitative result, especially in the energy of a molecule. There are several entries in this scheme such as MP, CI, CCSD(T), MCSCF, CASSCF, MRCI, and so on (see Sect. 3.1.2). The computation time required becomes much longer than the ordinary non-empirical HF calculation and much more memories for data storage becomes necessary at the same time. This scheme affords comparatively quantitative results so that one should consider the balance between the quality of the results and the computational resource when employing this scheme.

DFT scheme:

Recent progress and improvement of the DFT method is remarkable and one can obtain reasonably quantitative results out of this scheme. This has become rather popular with recent experimental chemists for interpretation of their experimental results as to the molecules and chemical reactions. The orbitals obtained are often mentioned as also MO's by the experimental users in spite of that strict theoreticians rigorously tend to quote those as Kohn-Sham (KS) orbitals rather than MO's.

There are several entries based on the hybrid functionals employed (B3LYP, B97, PBED, and so on) and on the semiempirical functionals that can include long-range corrections (LCBOP, CAM-B3LYP, ωB97X-D, and so on) (see Sect. 3.2). Among these, for instance, it looks like that the current experimentalists mostly use the B3LYP.

(2) Molecular mechanics (MM) method

This method is characterized by a molecular force field based on various kinds of potential functions employed. There have been developed several kinds of molecular force field targeting for various chemical systems. For instance, TIP4P force field is appropriate to the system including H_2O molecules as solvent, AMBER and CHARMM to biopolymer systems, and OPLS to liquid simulations. Moreover, MMFF94, MM2, MM3, MM4 force fields could be applicable to organic molecules as well as biopolymers (also see Sect. 3.4). Hence, the force field ought to be employed according to the chemical system one deals with.

Q2: What kind of basis set should be selected?

A2: It is crucial to employ an adequate basis set in the MO calculation. Basis set implies the collective group of the basis functions to construct the LCAO used in the MO calculation. In general, the more numbers of basis functions are included in a basis set, the better results especially as to molecular energy will be provided. In other words, such a basis set is referred to as that of higher quality but, as a matter of course, it requires the longer computation time. There have been hitherto proposed quite a few kinds of basis sets, which might be a bit ungraspable to ordinary users. Each basis set, however, has its own structure, features, and systematic naming. Hence, the users are encouraged to make out some of these and to select adequate one (ones) considering their calculation purpose. Those frequently used, within the Pople type (Ditchfield et al. 1971) as an example, are listed as follows:

(1) Minimal basis set

This is the least basis set and useful to qualitative analysis of the electronic structure of molecules. The least necessary orbitals are allotted for the inner-shell and valence-shell electrons in terms of Slater-type orbitals (STO's) (see Sect. 3.1.1). The minimal basis set is usually expressed as STO-NG ($N = 3, 4, 6$, etc.), where N signifies the number of the Gaussian-type orbitals (GTO's) to expand an STO.

The minimal basis set requires a comparatively short time for computation and it gives a pretty good result for the orbital patterns when considering the orbital interaction toward chemical reaction analysis. Actually, the orbital pattern pictures afforded by the minimal basis set are rather visible. The basis set consisting of too many basis functions is apt to result in over-delocalized MO's from the viewpoint of orbital patterns. Due to the same reason, the delocalization of electrons in the concerning molecule tends to be suppressed in the calculation using the minimal basis set, leading to rather rigid atomic net charges, for instance.

(2) Split-valence basis set

This is the basis set employing double or triple GTO sets with different orbital exponents to expand the STO's of the valence orbitals, which is called split valence. Different orbital exponent (also called zeta) results in different spatial extensions giving more quantitative results in energies compared with those by the minimal basis set. The split-valence basis set is usually expressed as N-KLG or N-KLMG for double zeta or triple zeta, case, respectively. The index N indicates the number of GTO's to expand the inner-shell orbital(s), and K, L or K, L, M the numbers of GTO's to expand the valence-shell orbitals for double zeta or triple zeta, respectively. Typical examples are: 3-21G, 4-31G, 6-31G (double zeta), and 6-311G (triple zeta).

Split-valence basis set gives more quantitative results in energies compared with those by the minimal basis set but necessitates more computation time.

(3) Polarized basis set

Addition of polarization functions (d or p functions) to the most outer orbital of the split-valence basis set. The d functions are added to the most outer p orbitals in the case of the atoms larger than Li atoms, and p functions for H atom. These d and p signs are added in the parentheses. Typical examples are: 6-31G(d), 6-31G(d, p), and 6-311G (d, p). The (d) or (d, p) is also indicated as * or **, respectively, as 6-31G* and 6-31G**.

A polarized basis set is good for the description of molecules having inner polarization since this can describe the polarization effect or deviation of electron distribution in the molecule. However, on one hand, the computation time becomes considerably longer with employing the polarized basis set.

(4) Diffuse basis set

Addition of diffuse functions is expressed by insertion of the sign + or ++ between split valence and G such as, e.g., 6-311 + G(d, p), 6-311 + G(2d, p), and 6-311 + +G(3df, 3pd), etc. Notations of diffuse basis sets are bit complicated, and implications of those are as follows:

6-311 + G(d, p): For the valence orbitals of atoms larger than Li, quadruple zeta of valence s and p functions and single zeta of d functions. For the valence orbital of H atom, triple zeta of s function and single zeta of p functions.

6-311 + G(2d, p): For the valence orbitals of atoms larger than Li, quadruple zeta of valence s and p functions and double zeta of d functions. For the valence orbital of H atom, triple zeta of s function and single zeta of p functions.

6-311 ++G(3df, 3pd): For the valence orbitals of atoms larger than Li, quadruple zeta of valence s and p functions, and triple zeta of d functions, and single zeta of f functions. For the valence orbital of H atom, quadruple zeta of s function, triple zeta of p functions, and single zeta of d functions.

*In the above, all the inner-shell orbitals are expanded by six GTO's.

Although the computation time largely increases by using a diffuse basis set, this kind of basis set well describes the electronic structure of anionic molecule, supermolecule, molecule with lone pairs, or rather high excited state of molecules, since electrons in these species are rather liberated from the original molecule.

Some others:

(5) cc-PVNZ; correlation consistent basis set (Dunning 1989)

(cc-P = correlation consistent polarized; V = valence orbitals only; N = D, T, Q, and so on, where D = double, T = triple, Q = quadruple; Z = zeta).

This basis set includes polarization functions from the first. Inclusion of diffuse orbitals is represented by putting prefix "aug-" like "aug-cc-PVDZ".

(6) Effective core potential (ECP)

This is not exactly the basis set for all the electrons particularly in heavy atoms whose inner core electrons are replaced with a certain potential form. There are several kinds of ECP's such as CEP-31G (compact effective potential plus 31G), LanL2DZ (Los Alamos national laboratory ECP plus DZ), SDD (Stuttgart/Dresden ECP), and so on. One can also use the ECP called RECP including the relativistic effect inherent in the heavy atoms.

Q3: Is there restrictions for the optimization of molecular structure?

A3: All the calculation methods cannot necessarily perform molecular structure optimization. For instance, the Hückel and the extended Hückel methods cannot perform the geometrical optimization, since they afford only the simple MO energies without inclusion of the interelectron interactions.

The optimization of molecular structure is possible by the usage of HF, post-HF, DFT, and MM methods. Broadly speaking, the post-HF or the DFT method can afford considerably precise molecular structures consisting of atoms of less than several hundreds. For large molecules such as polymeric, biological, and catalytic systems, MM method is rather effective than the elaborate MO methods. Care should be taken to the selection of the molecular force field in the MM depending on the target molecule. The ONIOM method (QM/MM) would also be applicable to those huge systems.

For the molecules in the excited state (S_1 or T_1), the CIS or the TD-DFT method can be applicable. Also, it is noted that, even in the ground state, anionic species ought to be dealt with the diffuse basis sets into the MO calculations.

Q4: Are there notes for the calculation of the electronic structure of a molecule?

A4: The MM method is not appropriate to obtain the electronic structure since this does not contain the information of electrons. Empirical MO and the semiempirical HF methods give only qualitative and/or partial results on the electronic structures but, on one hand, can supply considerably enough information for chemical reactivity analyses. For instance, early frontier orbital theory (Fukui et al. 1952, 1971) and Woodward-Hoffmann's rule (Woodward and Hoffmann 1965, 1969) came out based on the empirical MO methods. This comes from that the MO patterns and the order of MO energies are rather insensitive to the degree of the approximation employed in the MO theory. Incidentally, it is noted that in the CO patterns of the 1D polymers the graspable ones are only obtained at the center and the boundaries of the Brillouin zone (see Sect. 3.3) since the CO's is obtained as complex functions in other zone points.

More general electronic structures and properties such as atomic net charges, bond-order indices, dipole moment, polarizability, NMR shift (or NICS (see Sect. 2.8.2)), and so on should be enumerated by the non-empirical HF or the recent DFT method used in quantum chemistry. Care should be taken, however, that atomic net charges are rather sensitive to not only the MO method but selected basis sets due to the tendency of delocalization of electrons depending on those conditions as described in Sect. 2.3. It is mentioned that high quality of the basis set such as 6-311 + G(2d, p) or more should be employed to obtain a plausible result for the NMR shift.

Q5: Is it possible to get the IR vibration and Raman scattering data?

A5: One can obtain this kind of information out of the MM and the MO method, which can perform the optimization of molecular structure in **Q3**. This is because the normal vibration data can be obtained by the frequency analysis usually performed for the confirmation of the local minimum of the energy of the molecule. The vibration intensities are also enumerated if the electronic structure of the molecule is available and, in this sense, the MM method cannot afford these data. The zero-point frequency is also derived along with each normal vibration, as a matter of course, which is important for zero-point energy correction particularly in obtaining the thermochemical data concerning with **Q7**.

It is noted that, in particular, the frequencies tend to be overestimated by the usual MO method and a frequency scale factor is to be multiplied to the original data of all the normal vibration frequencies. A couple of instances are 0.8929 for HF/6-31G* data (Pople et al. 1993) and 0.989 for DFT/B3LYP/6-311 + G(3df, 2p) data (Bauschlicher and Partridge 1995), signifying that elaborate DFT generally gives better frequency values. For the general HF method, the above value (0.8929) is often used irrespective of the basis sets, and so for the general DFT method (0.989).

Q6: Are there notes for the calculation of excitation energies?

A6: These can be obtained either by MO or by DFT method with an assignment of the MO's concerning the electron transition. Although the usual HF method is capable of enumerating the excitation energy, usage of the configuration interaction (CI) or more quantitative coupled-cluster (CC) method is generally recommended since these methods include electron correlation effect, which is of importance in description of the excited state. In DFT method, the TD-DFT one has been developed for the calculation of excitation energies. Along with the excitation energy, the oscillator strength compared to the absorption coefficient is also obtained as has been described in Sect. 2.5.

Since the experimental absorption spectra are often measured in solution, it is required to include the solvation effect to the theoretically obtained data. For this purpose is usually utilized correction using the polarizable continuum model (PCM) for each solvent, in which the parameters of the conventional solvent species are included. An example of usage of the PCM has also been given in Sect. 2.5.

Q7: Can one obtain the thermochemical data such as enthalpy, entropy, and free energy of a molecule?

A7: It is possible that one obtains the data Gibbs free energy G, entropy S, and enthalpy H of a molecule after the optimization of molecular structure with frequency analysis of the normal vibrations. This is because the partition function including that of vibration is indispensable to the calculation of entropy of a molecule. Temperature T and pressure p are to be given as parameters for the actual calculation of these data. This process is not so tedious since frequency analysis is normally performed to carry on the structural optimization of a molecule. Moreover, the subroutine for the additional calculation of these thermochemical data is often equipped within the usual software of the MO calculation.

Q8: What is BSSE?

A8: In the calculation of supermolecule A . . . B such as a hydrogen-bonded system, for instance, one should pay attention to the basis set superposition error (BSSE). This occurs from that the moiety A automatically utilizes the basis set of the counter moiety B resulting in excessive stabilization of A, and vice versa to the moiety B. The BSSE especially tends to take place when the basis set functions of low quality are employed for the concerning molecule. This kind of error can be eliminated by the counterpoise method (Boys and Bernardi 1970).

Q9: How can TS and IRC be obtained?

A9: The procedure to search for the transition state (TS) is generally associated with that for the structural optimization of molecules. For TS finding is to examine the col (saddle point) having the local maximum of energy profile (on the potential surface) along the reaction coordinate assigned by the user. It is noted that, however, the TS has only one imaginary frequency of the normal vibration suggesting the direction of molecular deformation leading to the reaction product. This saddle point is equal to the TS. Hence, that at the saddle point the molecular vibration has only one imaginary frequency.

The potential energy barrier at the TS setting that of the reactant(s) as zero is normally estimated to be higher than that found by the experimental estimation. In this sense, one ought to select the calculation method that can afford the quantitative energy of the molecular species including also the TS state as much as possible.

The intrinsic reaction coordinate (IRC) is obtained as two routes: one is from the TS to the product(s) and another to the reactant(s) utilizing the formula defining the IRC on which the steepest descent path is guaranteed down from the TS. The actual calculation is performed at finite (discrete) numbers of points on each IRC, so one has to be careful not to stray off the intrinsic path.

References

C. Adamo, V. Barone, J. Chem. Phys. **110**, 6158–6170 (1999)
B.J. Alder, T.E. Wainwright, J. Chem. Phys. **31**, 459–466 (1959)
J. Antony, S. Grimme, Phys. Chem. Chem. Phys. **8**, 5287–5293 (2006)
C.W. Bauschlicher Jr., H. Partridge, J. Chem. Phys. **103**, 1788–1791 (1995)
A.D. Becke, Phys. Rev. A **38**, 3098–3100 (1988)
A.D. Becke, J. Chem. Phys. **98**, 5648–5652 (1993)
A. D. Becke, J. Chem. Phys. **107**, 8554-8560 (1997)
R.C. Bingham, M.J.S. Dewar, D.H. Lo, J. Am. Chem. Soc. **97**, 1285–1293 (1975)
S.F. Boys, F. Bernardi, Mol. Phys. **19**, 553–566 (1970)
J.-D. Chai, M. Head-Gordon, J. Chem. Phys. **128**(084106), 1–15 (2008a)
J.-D. Chai, M. Head-Gordon, Phys. Chem. Chem. Phys. **10**, 6615–6620 (2008b)
J. Čížek, J. Chem. Phys. **45**, 4256–4266 (1966)
A.J. Cohen, P. Mori-Sánchez, W. Yang, Chem. Rev. **112**, 289–320 (2012)
E.U. Condon, Phys. Rev. **36**, 1121–1133 (1930)
L.A. Curtiss, K. Raghavachari, G.W. Trucks, J.A. Pople, J. Chem. Phys. **94**, 7221–7230 (1991)
J. Del Bene, H.H. Jaffé, J. Chem. Phys. **50**, 1126–1129 (1969)
G. Del Re, J. Ladik, G. Biczó, Phys. Rev. **155**, 997–1003 (1967)
R. Ditchfield, W.J. Hehre, J.A. Pople, J. Chem. Phys. **54**, 724–728 (1971)
T.H. Dunning, J. Chem. Phys. **90**, 1007–1023 (1989)
J. Frenkel, *Wave Mechanics: Advanced General Theory* (Clarendon Press, Oxford, 1934)
D. Frenkel, B. Smit, *Understanding Molecular Simulation: From Algorithms to Applications*, 2nd edn. (Academic Press, San Diego, 2001)
K. Fukui, Acc. Chem. Res. **4**, 57–64 (1971)
K. Fukui, T. Yonezawa, H. Shingu, J. Chem. Phys. **20**, 722–725 (1952)
W.A. Goddard III, L.M. Harding, Ann. Rev. Phys. Chem. **29**, 363–396 (1978)

V. Heine, *Group Theory in Quantum Mechanics* (Pergamon Press, London, 1960)
J. Heyd, G.E. Scuseria, M. Ernzerhof, J. Chem. Phys. **118**, 8207–8215 (2003)
H. Hirao, K. Morokuma, J. Phys. Chem. Lett. **1**, 901–906 (2010)
R. Hoffmann, J. Chem. Phys. **39**, 1397–1412 (1963)
P. Hohenberg, W. Kohn, Phys. Rev. B **136**, 864–871 (1964)
E. Hückel, Z. Phys. **60**, 423–456 (1930)
H. Iikura, T. Tsuneda, T. Yanai, K. Hirao, J. Chem. Phys. **115**, 3540–3544 (2001)
A. Imamura, H. Fujita, J. Chem. Phys. **61**, 115–125 (1974)
C. Jamorski, M.E. Casida, D.R. Salahub, J. Chem. Phys. **104**, 5134–5147 (1996)
W.L. Jorgensen, J. Tirado-Rives, J. Comput. Chem. **26**, 1689–1700 (2005)
C. Kittel, *Introduction to Solid State Physics,* 8th edn. (Wiley, New York, 2004)
W. Kohn, L.J. Sham, Phys. Rev. A **140**, 1133–1138 (1965)
P.A. Kollman, I. Massova, C. Reyes, B. Kuhn, S. Huo, L. Chong, M. Lee, T. Lee, Y. Duan, W. Wang, O. Donini, P. Cieplak, J. Srinivasan, D.A. Case, T.E. Cheatham III, Acc. Chem. Res. **33**, 889–897 (2000)
J.B. Krieger, J. Chen, G.J. Iafrate, A. Savin, in *Electron Correlations and Materials Properties* ed. by A.G. Kioussis (Plenum, New York, 1999)
C. Lee, W. Yang, R.G. Parr, Phys. Rev. B **37**, 785–789 (1988)
P.-O. Löwdin, Phys. Rev. **97**, 1509–1520 (1955)
F. Maseras, K. Morokuma, J. Comput. Chem. **16**, 1170–1179 (1995)
N. Metropolis, A.W. Rosenbluth, M.N. Rosenbluth, A.H. Teller, E. Teller, J. Chem. Phys. **21**, 1087–1092 (1953)
C. Møller, M.S. Plesset, Phys. Rev. **46**, 618–622 (1934)
H. Nakatsuji, K. Hirao, J. Chem. Phys. **68**, 2053–2065 (1978)
H. Ohtani, Y. Hirao, A. Ito, K. Tanaka, O. Hatozaki, J. Therm. Anal. Calorim. **99**, 139–144 (2010)
R.G. Parr, W. Yang, *Density-Functional Theory of Atoms and Molecules* (Oxford Univ. Press, Oxford, 1989)
J.P. Perdew, Phys. Rev. B **33**, 8822–8824 (1986)
J.P. Perdew, Y. Wang, Phys. Rev. B **45**, 13244–13249 (1992)
J.P. Perdew, K. Burke, M. Ernzerhof, Phys. Rev. Lett. **77**, 3865–3868 (1996)
J.P. Perdew, S. Kurth, A. Zupan, P. Blaha, Phys. Rev. Lett. **82**, 2544–2547 (1999)
J.A. Pople, R.K. Nesbet, J. Chem. Phys. **22**, 571–572 (1954)
J.A. Pople, G.A. Segal, J. Chem. Phys. **44**, 3289–3296 (1966)
J.A. Pople, A.P. Scott, M.W. Wong, L. Radom, Israel J. Chem. **33**, 345–350 (1993)
B.O. Roos, P.R. Taylor, P.E.M. Siegbahn, Chem. Phys. **48**, 157–173 (1980)
C.C.J. Roothaan, Rev. Mod. Phys. **23**, 69–89 (1951)
C.C.J. Roothaan, Rev. Mod. Phys. **32**, 179–185 (1960)
J.R. Rumble, Jr., D.R. Lide, T.J. Bruno (eds.), *CRC Handbook of Chemistry and Physics*, 99th edn (CRC Press, Taylor & Francis Group, Boca Raton, Fl., 2018, pp. 9–65)
E. Runge, E.K.U. Gross, Phys. Rev. Lett. **52**, 997–1000 (1984)
O. Sinanoğlu, Adv. Chem. Phys. **6**, 315–412 (1964)
J.C. Slater, Phys. Rev. **81**, 385–390 (1951)
R. Stowasser, R. Hoffmann, J. Am. Chem. Soc. **121**, 3414–3420 (1999)
M. Svensson, S. Humbel, K. Morokuma, J. Chem. Phys. **105**, 3654–3661 (1996)
M. Svensson, S. Humbel, R.D.J. Froese, T. Matsubara, S. Sieber, K. Morokuma, J. Phys. Chem. **100**, 19357–19363 (1999)
J. Tao, J.P. Perdew, V.N. Staroverov, G.E. Scuseria, Phys. Rev. Lett. **91**(146401), 1–4 (2008)
T. van Voorhis, G.E. Scuseria, J. Chem. Phys. **109**, 400–410 (1998)
S.H. Vosko, L. Wilk, M. Nussair, Can. J. Phys. **58**, 1200–1211 (1980)
J.L. Whitten, M. Hackmeyer, J. Chem. Phys. **51**, 5584–5596 (1969)
R.B. Woodward, R. Hoffmann, J. Am. Chem. Soc. **87**, 395–397 (1965)
R.B. Woodward, R. Hoffmann, Angew. Chem. Intern. Ed. Engl. **8**, 781–853 (1969)
S. Xu, S. Irle, D.G. Musaev, M.C. Lin, J. Phys. Chem. A **109**, 9563–9572 (2005)

T. Yanai, D.P. Tew, N.C. Handy, Chem. Phys. Lett. **91**, 51–57 (2004)
S. Yanase, T. Oi, J. Nucl. Sci. Technol. **39**, 1060–1064 (2002)
R.-B. Zhang, X.-C. Ai, X.-K. Zhang, Q.-Y. Zhang, Solvent effects on the excited state properties of 2-aminopurine: a theoretical study by the ONIOM and supramolecular method, J. Mol. Struc. (Theochem) **680**, 21–27 (2004)

Chapter 4
Toward More Sophisticated Problems

Abstract In this chapter, examples of actual applications of theoretical chemistry toward several selected fields in chemistry are to be introduced. The fields dealt with here cover nanoscience, electrochemistry, catalytic chemistry, and theoretical biology. It will be hopefully understood that theoretical chemistry is readily useful in that it can afford us an interpretation of specific but interesting chemical phenomena, analyses of complicated chemical behaviors, and substantial molecular designs.

Keywords Application · Nanoscience · Electrochemistry · Catalytic chemistry · Theoretical biology

4.1 Nanoscience

It is rather recent that nanoscience became one of the foci of practical molecular science. In this field, molecular properties are attempted to directly connect to characteristics of nanoscale devices, and those properties are most effectively examined by theoretical chemistry at the molecular level. However, it should be noted that the essential methodology toward the theoretical analysis of nanoscience is still in progress. For instance, the interaction mode of cores of a nanoscale device with the environment such as the external electrodes has not yet been fully understood. In this section, a couple of examples on these subjects are to be given from the viewpoints of theoretical chemistry.

4.1.1 Connection of Molecular Wire to the External Electrode

Most of the molecular electronic devices necessitate the electric connection between the external electrodes and the core molecule of the device. This is possibly realized by injection and/or ejection of electrons through appropriate molecular wire provided with the anchor atoms, for which, e.g., –SH or –SiH group is used to form –S–Au or –Si–Au bond to the Au electrode as illustrated in Fig. 4.1. A (space) –C–Si anchor

K. Tanaka, *Theoretical Chemistry for Experimental Chemists*,
https://doi.org/10.1007/978-981-15-7195-4_4

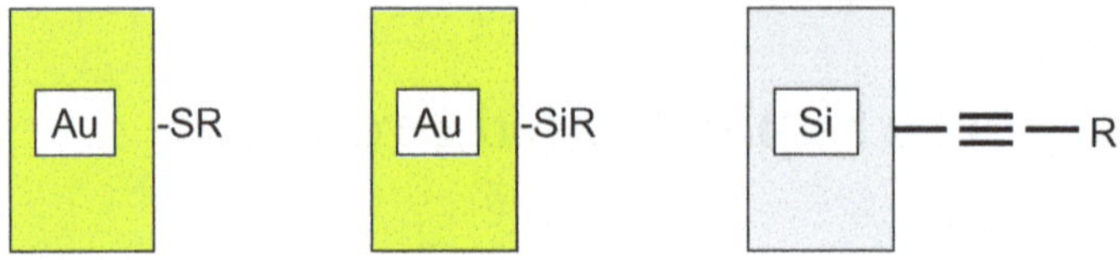

Fig. 4.1 Schematic drawings of connection of molecular wire R via various anchors (S, Si, and C) to the external bulk electrode (Au or Si)

bound to the Si electrode gives another possibility. The molecular wire R is further connected to the core molecule of nanodevices through a certain chemical bond. Among those, the most popular connection has been established by –SH anchor to the Au electrode (Lindstrom et al. 2005). The valence electrons of Au atom are in the $5d^{10}6s^1$ configuration and the Fermi level of Au bulk consists of 6s band. The chemical bond picture for –S–Au linkage has not necessarily become clear. For instance, it is still controversial which of 6s or 5d AO's mainly contributes to the Au–S bond (Tachibana et al. 2002; Di Felice et al. 2003).

In this section, a theoretical attempt to clarify this point using the concept of natural bond orbital (NBO) is to be introduced. To represent the Au electrode, a cluster model consisting of three layers of totally Au_{37} atoms in Fig. 4.2a for the (111) surface has been employed and a tetrathiafulvalene (TTF) dithiolate derivative (TTF-$(CH_2SH)_2$) in Fig. 4.2b adopted for the R-SH molecule. The Hartree-Fock (HF) calculation method was used for the total system with the basis sets Lanl 2 MB for Au, 3-21 + G* for the TTF molecule, and 3-21G* for CH_2S moieties with the diffuse function of S atoms (see Sect. 3.6 for the basis set). The interatomic distance between Au atoms was kept as 2.884 Å as in the bulk structure (Rumble 2018). All the interatomic distances in the TTF dithiolate derivative and those between this molecule and the modeled Au (111) cluster have been structurally optimized. There have been found out at least three kinds of possible binding structures for this TTF dithiolate derivative onto the Au (111) surface as shown in Fig. 4.3, where two S atoms are bound to the bridge sites (Fig. 4.3a) and the bridge and the atop sites

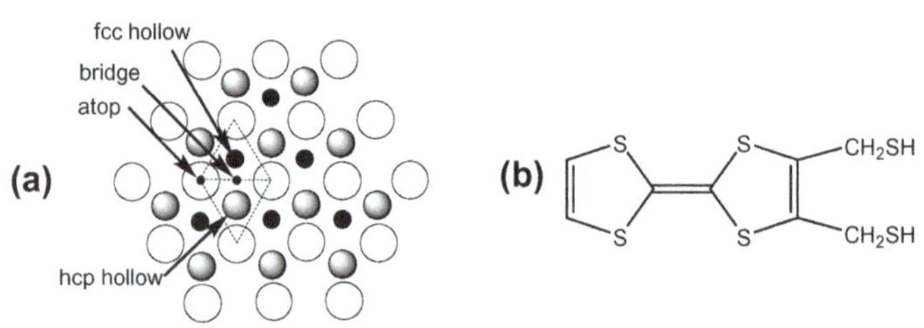

Fig. 4.2 **a** Au (111) surface model cluster consisting of 37 atoms with the indication of four kinds of possible binding sites: white, gray, and black balls are on the first, second, and third layers, respectively, and **b** tetrathiafulvalene (TTF) dithiolate derivative as a model of molecular wire having two S-anchors. Reprinted from Fueno et al. (2006). Copyright 2006, with permission from Elsevier

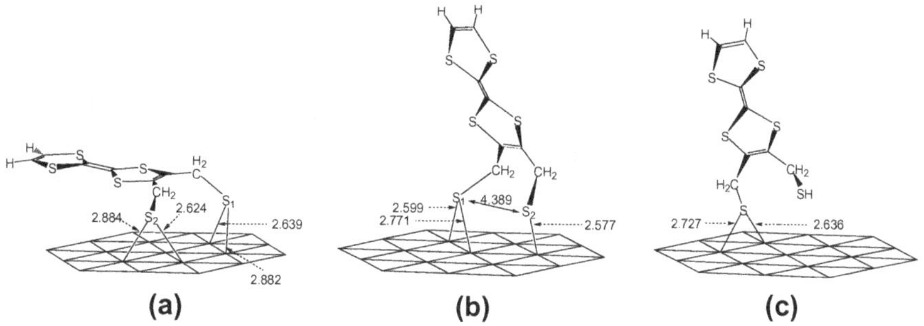

Fig. 4.3 Three kinds of the optimized structures of TTF-dithiolate derivative bound to the Au (111) surface in Fig. 4.2a. The binding energies of these are **a** 113.14 kcal/mol, **b** 109.89 kcal/mol, and **c** 66.84 kcal/mol. The figures indicate interatomic distance in Å. See text as to the calculation basis set. Reprinted from Fueno et al. (2006). Copyright 2006, with permission from Elsevier

(Fig. 4.3b) with the binding energies of about 110 kcal/mol. The binding energy in the structure where only one S atom is bound (Fig. 4.3c) is approximately half of those.

In order to clarify the nature of the Au–S bond, the NBO analysis was performed as to the binding structure in Fig. 4.3b for instance. In the NBO analysis, the canonical MO's are transformed into an equivalent set of localized bond orbitals so as to clearly figure out the direct picture of the concerning bonds as has been described in Sect. 2.2 (Reed et al. 1988). In Table 4.1, the NBO analysis results for the binding structure in Fig. 4.3b are listed. It is clear that the Au 6 s AO contributes to the Au–S bonding orbital as well as to the antibonding orbital rather than 5d AO's. These NBO patterns thus obtained are shown in Fig. 4.4 including those originating from the σ and π lone

Table 4.1 Natural bond orbital (NBO) analysis[1] for the binding structure shown in Fig. 4.3b

Orbital pattern type[2]	Electron occupancy	Orbital energy (eV)	Orbital nature[3]	Participation ratio of each atom (%)	Contribution ratio of each AO (%)		
					s	p	d
(a)	1.00	−16.80	σ lone pair of S	–	73.01	26.97	0.02
(b)	0.95	−8.32	Au–S bonding	Au: 15.07	95.58	2.25	2.16
				S: 84.93	10.46	89.41	0.13
(c)	0.98	−4.33	π lone pair of S	–	0.03	99.95	0.01
(d)	0.27	3.82	Au–S antibonding	Au: 84.93	95.58	2.25	2.16
				S: 15.07	10.46	89.41	0.13

Reprinted from Fueno et al. (2006). Copyright 2006, with permission from Elsevier

[1]See text as to the calculation basis set

[2]See Fig. 4.4

[3]All the S atoms in this Table signify S_2 bound to the atop site in Fig. 4.3b

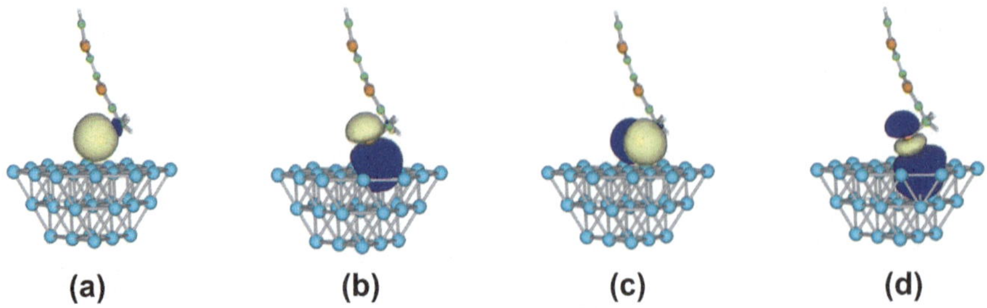

Fig. 4.4 NBO patterns of **a** σ lone pair of S, **b** Au–S bonding orbital, **c** π lone pair of S, and **d** Au–S antibonding orbital in the binding structure in Fig. 4.3b. Reprinted from Fueno et al. (2006). Copyright 2006, with permission from Elsevier

pairs of the S atom.

4.1.2 Molecular Field-Effect Transistor (FET)

In molecular nanotechnology, it is often required to estimate the current-voltage ($I-V$) characteristics of electrically conductive molecular wire connecting between two nanoelectrodes. A more sophisticated molecular wire is equipped with a quantum dot in its middle area separated by energy barriers from the wire portions as illustrated in Fig. 4.5. This kind of fabrication is of importance to design molecular field-effect transistor (FET). In this subsection, theoretical examination of electric behavior is to be described by consideration of design toward molecular wire model with a quantum dot consisting of single-walled carbon nanotube (SWCNT) of chiral indices (6, 6) with a finite length as an example, theoretical examination of its electric behavior is to be described.

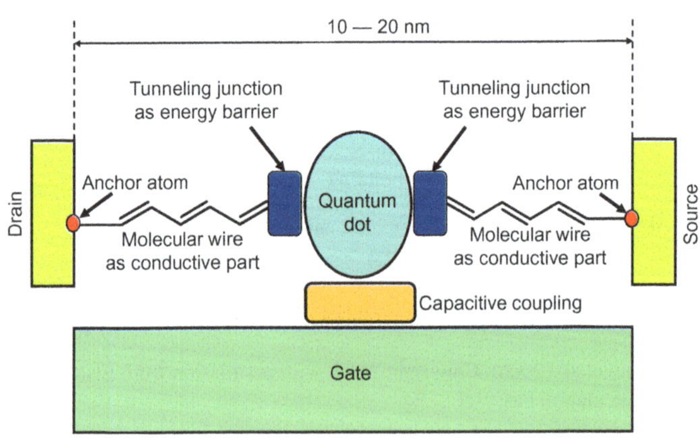

Fig. 4.5 Schematic drawing of a molecular field-effect transistor (FET) device

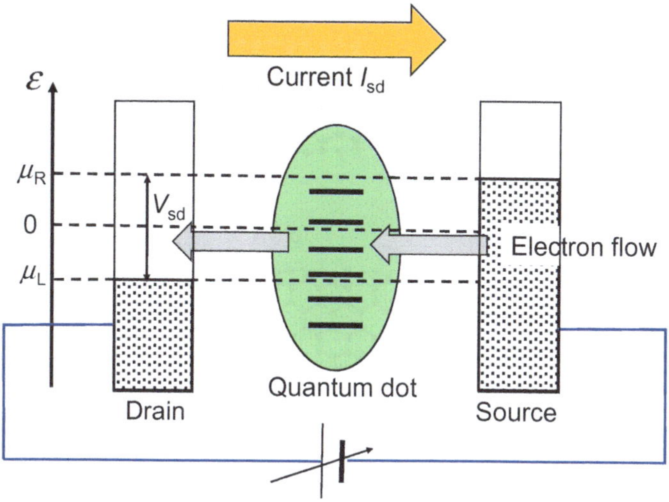

Fig. 4.6 Quantum dot for a molecular FET system

The calculation for this purpose requires an additional scheme to the ordinary HF or DFT methodology, which is called non-equilibrium Green's function (NEGF) scheme (Datta 2005). This eventually enables the numeration of the electric current I_{sd} as a function of the voltage applied between the source and drain electrodes V_{sd} shown in Fig. 4.6 by

$$I(V) = G_0 \int_{-\infty}^{\infty} d\varepsilon \{[n_F(\varepsilon - \mu_L) - n_F(\varepsilon - \mu_R)] \times \mathbf{T}(\varepsilon)\} \tag{4.1}$$

where ε stands for the energy, $n_F(\varepsilon)$ the Fermi distribution function, and the conductance quantum G_0 is expressed by

$$G_0 \equiv \frac{2e^2}{h} \tag{4.2}$$

with h and e representing the Planck constant and the elementary electric charge, respectively. The chemical potentials of the right (source) and the left (drain) electrodes are denoted as μ_R and μ_L, respectively, and $\mathbf{T}(\varepsilon)$ the transmission probability expressing electron drift ratio from the source to the drain electrode throughout the molecular wire including the quantum dot. The potential width $\mu_R - \mu_L$ is equal to V_{sd} also called bias voltage. Actual integration in Eq. (4.1) is performed within an appropriate range of ε including μ_L and μ_R. Thorough prescription of the computational work in the above has been given elsewhere (Stokbro et al. 2005).

The examination result with respect to the molecular wire model using CNT in Fig. 4.7a is explained in the below to afford an idea of these kinds of calculations

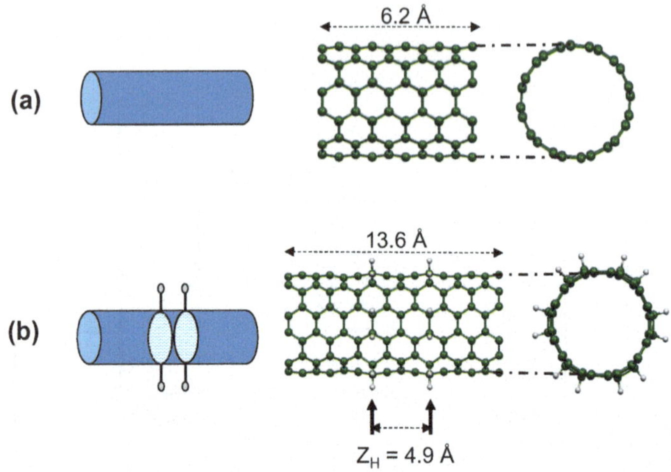

Fig. 4.7 **a** Carbon nanotube (CNT) with finite length (6.2 Å) considered here and **b** H-CNT (length 13.6 Å) modified with 24 hydrogen atoms (pointed by arrows) in total around its surface to yield energy barriers. For the source and the drain, electrodes shall be also used CNT with semi-infinite lengths. Adapted by permission from Fueno et al. (2012). Copyright (2012)

(Fueno et al. 2012). The energy barrier to separate the quantum dot is done by the distribution of hydrogen atoms around the surface of CNT as shown in Figs. 4.7b, 4.8. The computation has been carried out with the use of a commercial software TranSIESTA-C 1.3 program package (Taylor et al. 2001; Brandbyge et al. 2002)

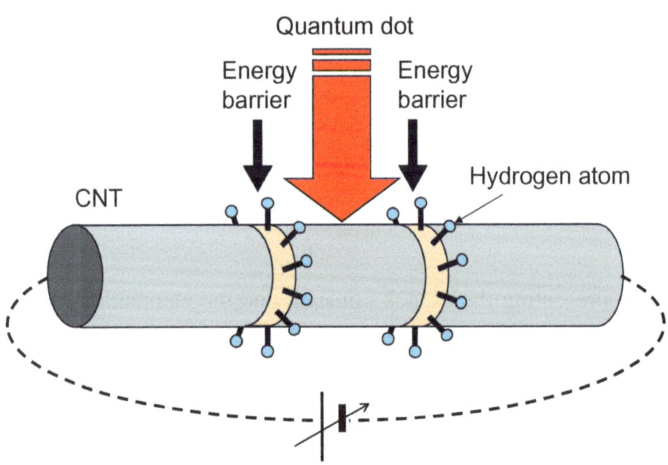

Fig. 4.8 Illustrative image of molecular FET using H-CNT in Fig. 4.7b. The introduction of two rings with C–H bonds makes energy barriers with sp^3 carbons between which a quantum dot appears

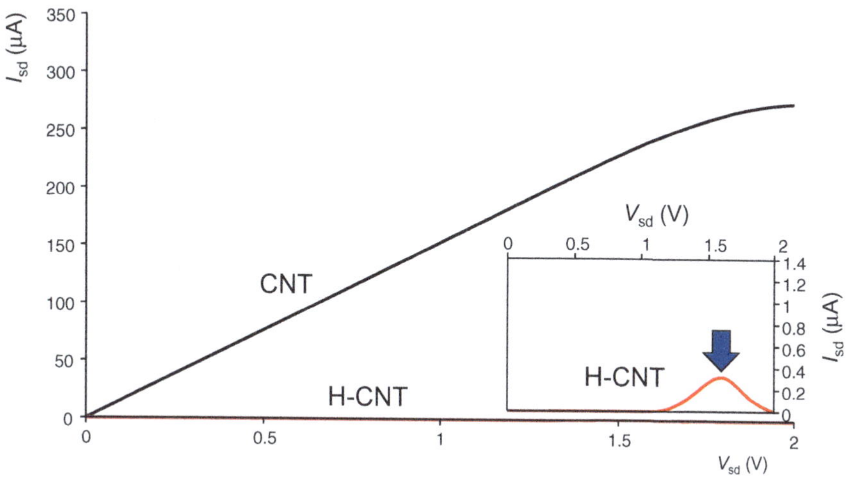

Fig. 4.9 $I_{sd}-V_{sd}$ characteristics of CNT (black) and H-CNT (red). The inset shows that of H-CNT with enlargement of I_{sd}. Adapted by permission from Springer Nature: Springer Nature, *Science China Chemistry*, Theoretical study of current-voltage characteristics of carbon nanotube wire functionalized with hydrogen atoms, H. Fueno, Y. Kobayashi, and K. Tanaka, Copyright (2012)

essentially based on the Perdew-Zunger local density approximation parameterization (Perdew and Zunger 1981) (LDA-PZ) with the single-zeta basis function (Soler et al. 2002). There are some features in the calculation result as follows:

(1) Calculated $I_{sd}-V_{sd}$ characteristics of CNT and H-CNT (inset) are shown in Fig. 4.9. There are three remarkable features: (i) The ohmic behavior is seen up to 1.6 V for CNT, a current peak pointed by an arrow appears at 1.6 V for H-CNT accompanied by the succeeding negative differential resistance (NDR) behavior signifying that current falls down with an increase in the applied voltage; (ii) the $I_{sd}-V_{sd}$ curve saturates at about 1.8 V for CNT; and (iii) the current in H-CNT is in the order of one-hundredth of that of CNT for V_{sd} ranging from 1 to 2 V.
(2) The transmission spectra of CNT and H-CNT with different V_{sd} are shown in Fig. 4.10. The energy ε corresponds to the ordinate in Fig. 4.6. The origin of energy ε is the average chemical potentials (or Fermi energies) of μ_L and μ_R as described above. The transmission spectra area surrounded by dashed triangular lines corresponds to those within the bias window. For CNT all the transmission spectra remain 2 in the bias-window range up to 1.6 V where the $I_{sd}-V_{sd}$ characteristics become ohmic since the bias window width is proportional to V_{sd}. This suggests that CNT behaves like a good molecular wire in itself. On the other hand, these spectra for H-CNT generally remain smaller than 0.2 except for several sharp peaks. In particular, there is quite a small peak at $V_{sd} = 1.6$ V in the range of the bias window. This small peak corresponds to the current peak at $V_{sd} =$ ca. 1.6 V in the inset of Fig. 4.9 at which the NDR sets in.
(3) The unoccupied orbitals of the molecular projected self-consistent Hamiltonian (MPSH) (Taylor et al. 2001; Brandbyge et al. 2002) of CNT shown in Fig. 4.11a,

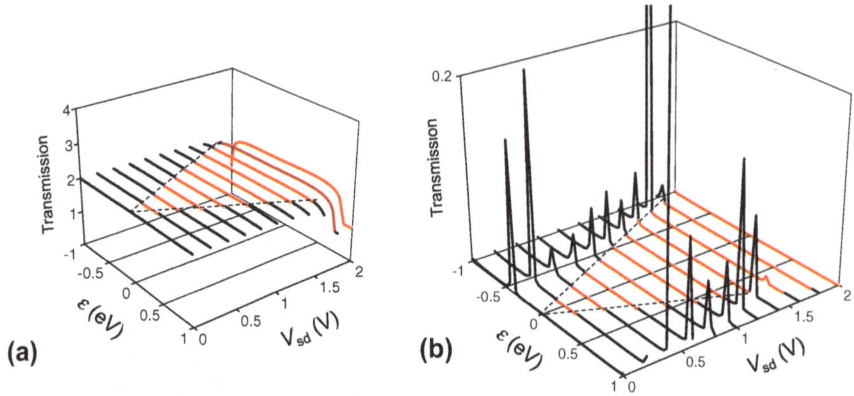

Fig. 4.10 Transmission spectra $T(\varepsilon)$ of **a** CNT and **b** H-CNT. The triangular area shown in red between the broken lines indicates the bias window. Adapted by permission from Fueno et al. (2012). Copyright (2012)

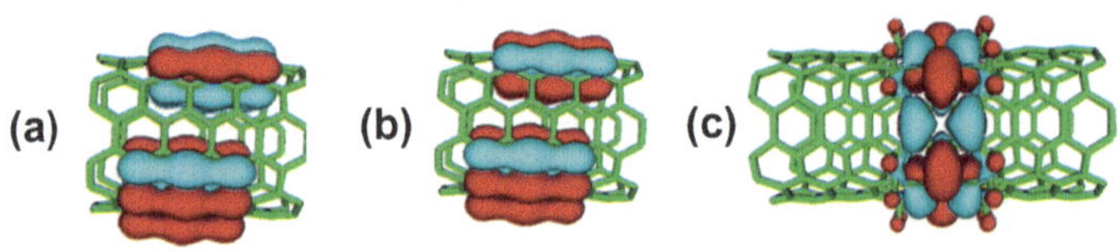

Fig. 4.11 Unoccupied orbital patterns with the energies of **a** −0.01 and **b** 0.01 eV contributing I_{sd} at $V_{sd} = 0.2$ V for CNT and of **c** 0.63 eV at $V_{sd} = 1.6$ V for H-CNT. Adapted by permission from Fueno et al. (2012). Copyright (2012)

b have the orbital energies of −0.01 and 0.01 eV, respectively. These are delocalized over the central region and contribute to electric current at $V_{sd} = 0.2$ V. On the other hand, as to H-CNT, it is seen in Fig. 4.11c that the unoccupied orbital whose energy is 0.63 eV is confined in the Z_H region under the condition of $V_{sd} = 1.6$ V. It is thus understood that the Z_H region behaves as a quantum well for H-CNT.

(4) Fig. 4.12 shows the gate voltage (V_g) dependencies of $I_{sd}-V_{sd}$ characteristics. Note that the orbital energy level in the quantum dot region can be controlled by applying V_g. It is seen that in Fig. 4.12a for CNT the change in V_g does not give remarkable effect to the $I_{sd}-V_{sd}$ characteristics, since there is no quantum dot in the simple CNT. On the other hand for H-CNT, the $I_{sd}-V_{sd}$ characteristics show typical oscillation tendency and increase in the number of peaks upon the change in V_g as seen in Fig. 4.12b. This obviously signifies the resonant tunneling in which the orbital energy levels are made to match with those of the outer wire portions by changing V_g.

It should be pointed out that I_{sd} in CNT is obviously large (100 μA order) in spite of a single molecular wire, which comes from the overestimation of the transmission

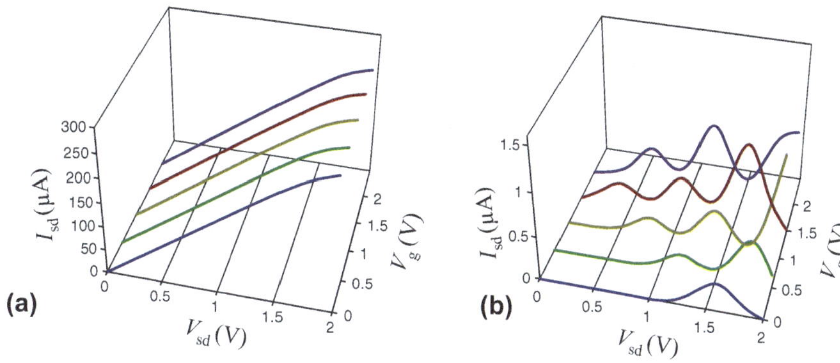

Fig. 4.12 V_g dependencies of the I_{sd}-V_{sd} characteristics of **a** CNT and **b** H-CNT. Adapted by permission from Fueno et al. (2012). Copyright (2012)

spectrum obtained here. Improvement of this situation could be brought about by consideration of the TD-DFT scheme (Koentopp et al. 2008).

4.1.3 Molecular Design Toward Nanospin Device

High-spin organic molecules are of interest per se and have been eagerly developed toward making, e.g., "molecular" magnet. Part of fundamental molecular design along this line has already been described in Sect. 2.8. It is, however, almost clear at present that the prospect of molecular magnet has some difficulty to be realized since the total antiferromagnetic effect in the bulk could cancel out the ferromagnetic correlation. Instead, it is now anticipated to fabricate core parts in the nanospintronics field rather than previous attempts of design toward molecular magnets. Relating to this topic an example of molecular design toward two-dimensional (2D) high-spin organic polymers is to be afforded below.

The basic idea for 2D high-spin organic polymers lies in utilizing the combination of alternating "meta" and "para" phenylene linkages both including nitrogen atoms which can be oxidized to become a nitrogen cation radical as illustrated in Fig. 4.13a, b (Ito et al. 2000). It is well known that the meta-linkage of phenylene is generally effective for parallel alignment of the electronic spins by spin polarization, which is expected to generate ferromagnetic property (Mataga 1968). Moreover, it has been clarified that the para-linkage is efficient for stabilization of spins in terms of the delocalization of spins to a certain appropriate spatial extent as seen in Würster's blue-based di(cation radical) (Ito et al. 1999). Generation of parallel spins in Fig. 4.13a is well explained by the utilization of non-bonding MO (NBMO) due to degeneracy or pseudo-degeneracy of the singly occupied MO (SOMO) levels appearing based on the Hund's rule in meta phenylene linkage as shown in Fig. 4.13c. Thus, the combination of meta-para linkages is expected for stable ferromagnetic correlations of spins in

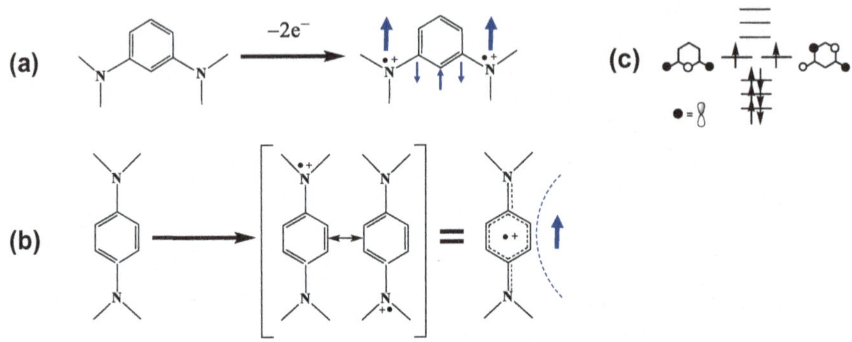

Fig. 4.13 **a** Generation of ferromagnetic spin correlation of *m*-phenylene diamine (MPD) dication radical by two-electron oxidation, **b** stabilization of a spin in *p*-phenylenediamine (PPD) cation radical, and **c** MO diagram of MPD with the generation of the two parallel spins in two degenerated NBMO's

total. In fact, it has been demonstrated that an actually synthesized molecule with three meta-para motifs shown in Fig. 4.14 has been proved to have quartet-spin state ($2S + 1 = 4$) when it is oxidized up to a trication (Ito et al. 2008). Moreover, an oligoarylamine molecule with more extended meta-para motifs as shown in Fig. 4.15 has been observed to reveal the septet state when it is oxidized up to hexacationic state (Sakamaki et al. 2009).

Degeneracy of the SOMO levels in the above would cause a super-degenerated band being half-filled, which works as the spin supplier in the polymer containing meta-phenylene linkages as illustrated in Fig. 4.16. An extended molecular design using this linkage toward 2D polymers as shown in Fig. 4.17 have been attempted and

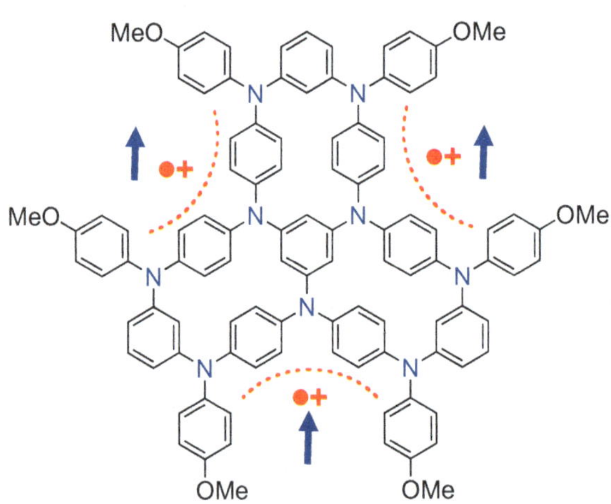

Fig. 4.14 Generation of stable quartet spin radicals in terms of the combination of meta and para (m, p) PD linkages

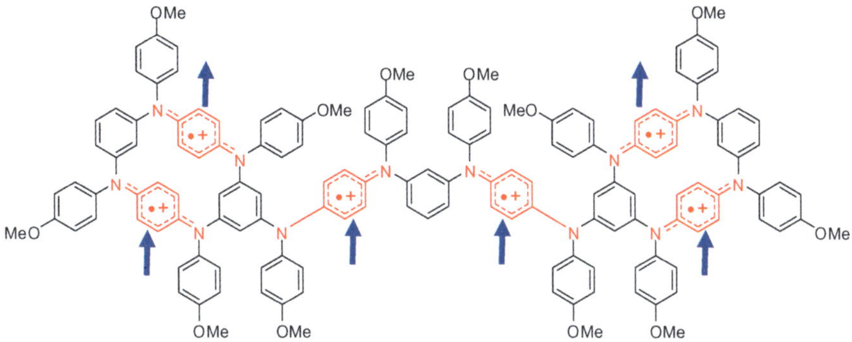

Fig. 4.15 Generation of stable septet spin radicals in terms of the combination of meta and para (*m*, *p*) PD linkages. The para motifs are indicated by red

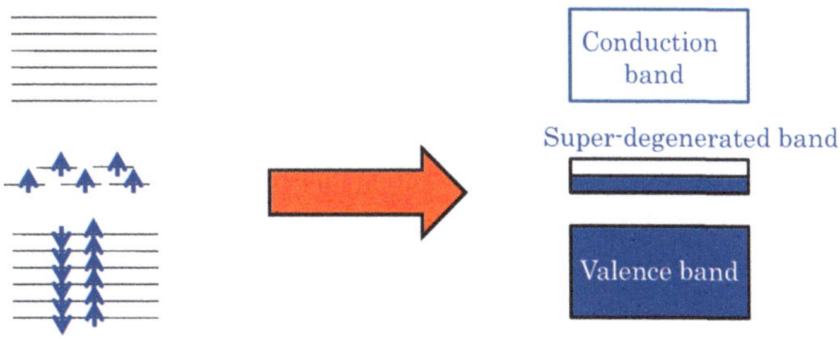

Fig. 4.16 Several degenerated or pseudo-degenerated NBMO's resulting in parallel spins based on the Hund's rule (left). This concept could be extended to polymers to possibly make a super-degenerated band supplying parallel spins (right)

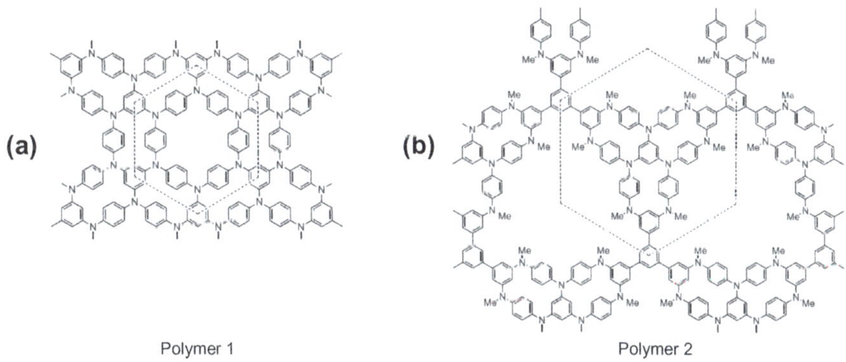

Fig. 4.17 Two kinds of unit cells enclosed by broken lines for molecular design toward 2D high-spin polymers **a** 1 and **b** 2. Reprinted from Ito et at. 2009, Copyright 2009, with permission from Elsevier

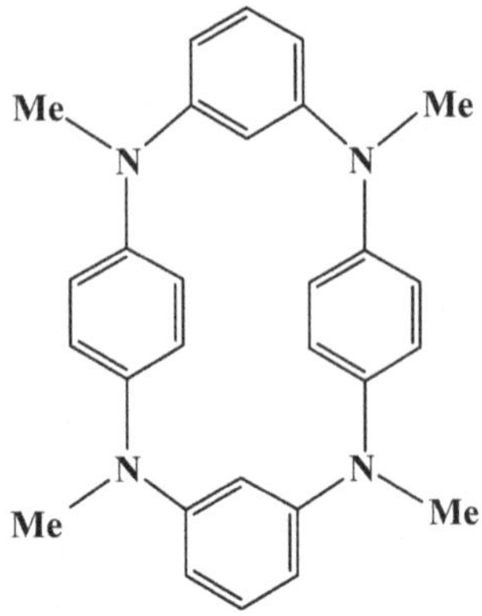

Fig. 4.18 Tetraaza[1.1.1.1]*m,p,m,p*-cyclophane. This motif includes meta- and para-phenylene linkages and is the essential part of the unit cells of 2D high-spin polymers 1 and 2 (see Fig. 4.17)

examined based on the theoretical calculation for polymers (see Sect. 3.3) in what follows (Ito et al. 2009). Both of the two kinds of unit cells **1** and **2** in Fig. 4.17a, b contain a combination of meta- and para-linkages composed of nitrogen atoms. Since the theoretical calculations of these 2D polymers consisting of these unit cells impose rather time-consuming computations, the molecules corresponding to the unit cells were first structurally optimized using a semiempirical MO calculation (AM1) and just one-point 2D-crystal orbital (CO) calculations for the neutral as well as for the multicationic unit cells ($\mathbf{1}^{3+}$, $\mathbf{2}^{3+}$, $\mathbf{1}^{6+}$, and $\mathbf{2}^{6+}$) were performed by employing the unit cells constructed from the above molecules without essential structural change. This is because the essential part (tetraaza[1.1.1.1]*m,p,m,p*-cyclophane in Fig. 4.18) of the molecules corresponding to the unit cells of **1** and **2** does not change the structure between the neutral and the higher cationic state by the DFT calculations (Ito et al. 2000).

The 2D polymer calculations were based on the unrestricted HF (UHF)-CO calculation with the 6-21G basis set. Characteristic molecular structures obtained by the AM1 calculations described above well reproduce those of arylamine molecules obtained by the X-ray crystallographic observations (Ito et al. 2000; Hauck et al. 1999) claiming that all the N atoms lie on a plane and that 1,3,5-benzenetriyl rings are on the same plane, from which the para-phenylene rings are canted out of the plane by 40–80°. The above AM1 calculations for the molecules representing the unit cells of **1** and **2** actually concluded the canting angles 59.0 and 47.2°, respectively. The 2D-CO calculations predicted that both **1** and **2** with all the multicationic unit cells ($\mathbf{1}^{3+}$, $\mathbf{2}^{3+}$, $\mathbf{1}^{6+}$, and $\mathbf{2}^{6+}$) favor the high-spin states as listed in Table 4.2 probably because the combination of meta- and para-linkages works well in those 2D polymers. This material could open up a possibility of high-spin nanosheet with appropriate synthetic preparation.

Table 4.2 Comparison of energies[1] of neutral and cationic polymers 1 and 2 in Fig. 4.17

Polymer	Charge per unit cell	$\Delta E_{L\text{-}H}$ [2]
1	Neutral	–
	+3	3.1
	+6	11.8
2	Neutral	–
	+3	3.7
	+6	19.6

[1]Based on the HF and UHF/CO/6-21G calculations
[2]In eV: Positive value signifies the stability of the higher spin state

4.2 Electrochemistry

The electrochemical process is of much importance both scientifically and technologically. In the electrochemical process, the central idea is based on the redox behaviors of materials and the electron transfer accompanied by this process. For a considerably long time, however, ambiguity had been remaining in many phenomena as far as starting from the conventional understanding through classical electrochemistry. This is partly because that the electrochemical process is not based on simple redox process but consists of a collection of various complicated interfacial problems such as those between the electrodes and the medium of electrons or ions, actual shape and behavior of ions in the electrolyte solution, actual pictures for ion migration, effective electron transfer, participation of irradiated light, and many other factors. Hence the attempts toward the total understanding of these ought to encounter various difficulties. Fortunately, in recent years, various observation techniques such as Raman spectroscopy, photoemission spectroscopy (PES), atomic force microscopy (AFM), soft X-ray emission spectroscopy (SXES), and so on have been developed to directly check the electrochemical events. The theoretical calculation could also become one of these to assist the understandings of some complicated events involved in electrochemical processes in many ways.

In this section, a couple of examples studied by theoretical calculations are elucidated to help understanding how it works in the actual examples in electrochemistry.

4.2.1 Solvation of Metallic Cation

It is well known that, in the rechargeable lithium-ion battery (LIB), Li^+ ions migrate from the positive electrode to the negative electrode in the charging process, and to the opposite direction in the discharging process through the electrolyte solution consisting of organic solvent and supporting electrolyte as illustrated in Fig. 4.19.

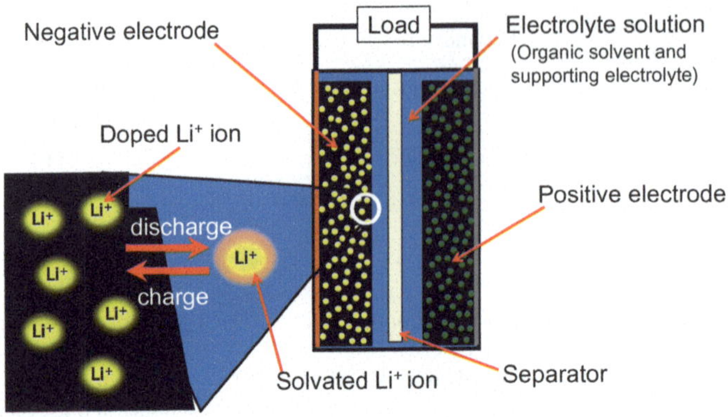

Fig. 4.19 Fabrication of lithium-ion battery (LIB) and an image of solvated Li^+

Although the overall behaviors of LIB can be analyzed by electrochemical measurements, theoretical calculation possibly affords the aid in examination and understanding of detailed behavior of electrons and Li^+ in the electrolyte solution and near the electrode at the molecular level. This is because a plenty of molecular processes are involved in the redox processes of the rechargeable of LIB.

One of the intriguing points would be the behavior of Li^+ ions in the electrolyte solution. For instance, Li^+ in the electrolyte solution is considered to cause solvation surrounded with several organic solvent molecules such as ethylene carbonate (EC) and propylene carbonate (PC) shown in Fig. 4.20. Often used as the supporting electrolytes are $LiPF_6$, $LiBF_4$, or $LiClO_4$. The theoretical calculation provides the optimized structures of solvated Li^+ with solvent molecules. For instance, the DFT/B3LYP/6-31G** calculations show the maximum number of the solvent molecules participating in solvation is four as to EC and PC with an approximate dimension of the spheres of solvations as seen in Fig. 4.21a, b, in which the solvent molecules surround the central Li^+ in a tetrahedral manner (Kaji et al. 2011). That the maximum solvation number of Li^+ is four has also been theoretically examined (Yanase and Oi 2002; Ohtani et al. 2010). The $Li^+(EC)_4$, for instance, the highest occupied MO (HOMO) and the lowest unoccupied MO (LUMO) are both almost triply degenerate since its structure is a little bit off from the T_d symmetry due to deformation caused by orientation of solvent molecules. The patterns of the HOMO and the LUMO are given in Fig. 4.22. Since both of these are one of the triply

Fig. 4.20 Typical solvent molecules for the electrolyte solution. **a** Ethylene carbonate (EC) and **b** propylene carbonate (PC)

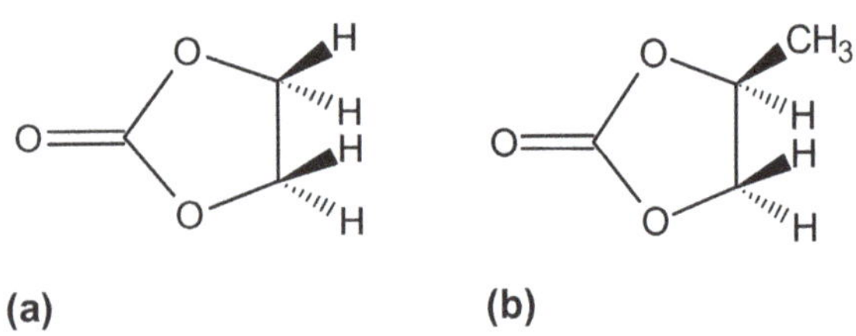

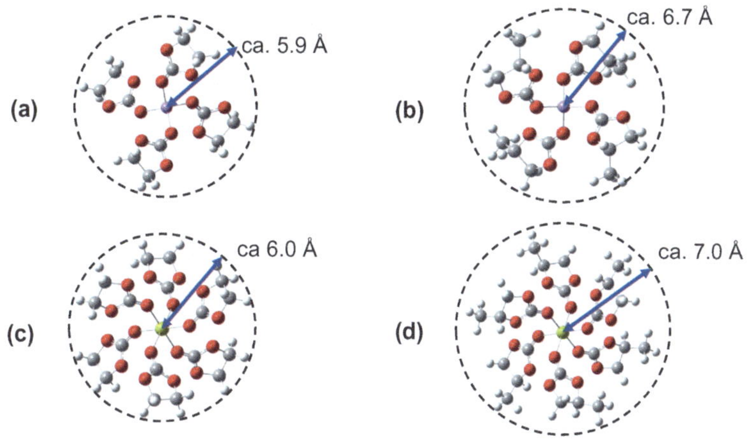

Fig. 4.21 Optimized structures of solvated Li^+ and Mg^{2+}. **a** $Li^+(EC)_4$, **b** $Li^+(PC)_4$, **c** $Mg^{2+}(EC)_6$, and **d** $Mg^{2+}(PC)_6$

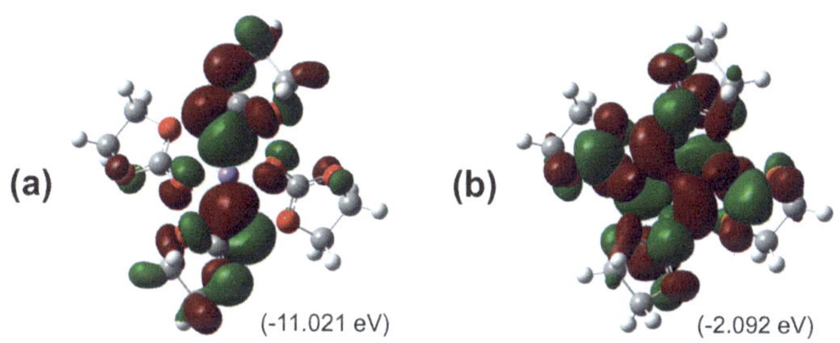

Fig. 4.22 Patterns and energies of Li + $(EC)_4$: **a** HOMO and **b** LUMO. Note that energies of the HOMO-1 and the HOMO-2 are −11.130 eV and −11.147 eV, respectively, and that those of the LUMO + 1 and the LUMO + 2 are −2.013 eV and −2.004 eV, respectively, implying pseudo-triple degeneracies of the HOMO and the LUMO

pseudo-degenerated MO's, their patterns are slightly deviated in a partial portion of the solvated species. Moreover, it has also been revealed (Kaji et al. 2011) that the solvated Mg^{2+} is surrounded by six solvent molecules at a maximum in an octahedral manner as shown in Fig. 4.21c, d, where Mg^{2+} ions are simply expected to carry charge twice as much as Li^+.

It is of interest that these solvated metallic cations should be desolvated upon the charging process near the negative electrode. To represent the carbon negative electrode, two kinds of simple models using nanographene(s) are employed as illustrated in Fig. 4.23. The enthalpy changes according to the stepwise desolvation process near the modeled carbon electrode have been examined with respect to $Li^+(EC)_4$ and $Mg^{2+}(EC)_6$ as represented in Figs. 4.24 and 4.25, respectively. It is seen that the desolvation enthalpies gradually become larger with the degree of the desolvation process and become the largest for desolvation of the final solvent molecule. These

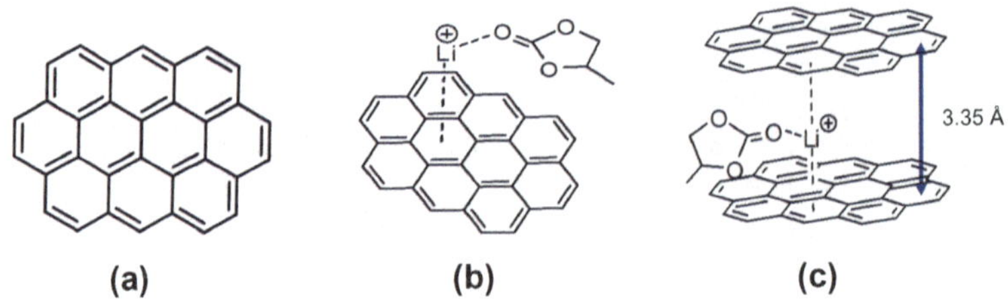

Fig. 4.23 Models for desolvation of $Li^+(PC)_1$ near the carbon electrode: **a** Nanographene (ovalene; $C_{32}H_{14}$) as a model of the electrode, and desolvation, **b** on the basal surface, and **c** at the edge surface

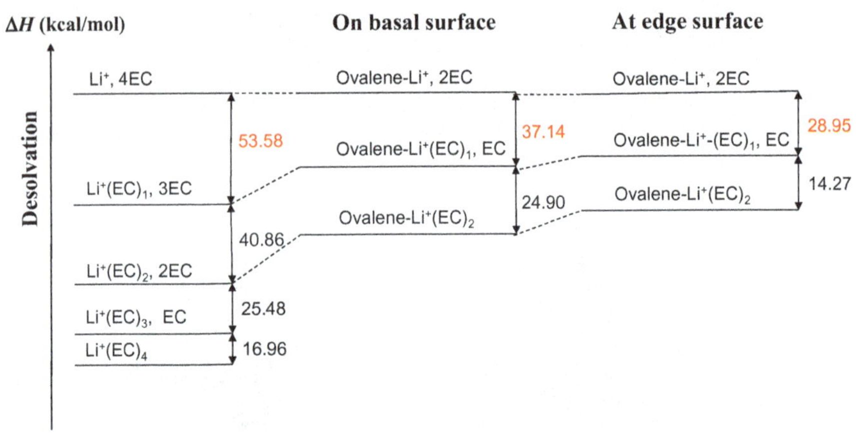

Fig. 4.24 Stepwise desolvation diagram of $Li^+(EC)_4$ with the enthalpy change (at 298.15 K and 1 atom) accompanied. The left represents the desolvation without the carbon electrode (see Fig. 4.23)

values, however, are considerably reduced on the carbon electrode compared with the carbon-free case for both the solvated Li^+ and Mg^{2+}. Moreover, it is clearly seen that desolvation enthalpies of Mg^{2+} are much larger compared with those of Li^+ implying that larger stabilization occurs in the solvation of Mg^{2+}.

4.2.2 *Surface of the Negative Electrode of the LIB*

One of the incomprehensible problems in LIB would be the formation of solid electrode interphase (SEI), being a kind of passivation layer, on the carbon negative electrode during the charging-discharging process. In order to address this situation, a plenty of researches have been piled up with respect to the formation and role of SEI (Winter 2009; Xu and von Cresce 2011; Tasaki et al. 2011; Xing et al. 2012). Here is introduced an attempt of those by theoretical calculation using the DFT/B3LYP/6-311++G* method (Xing et al. 2018), in which the electrolyte $LiPF_6$ in an LIB is

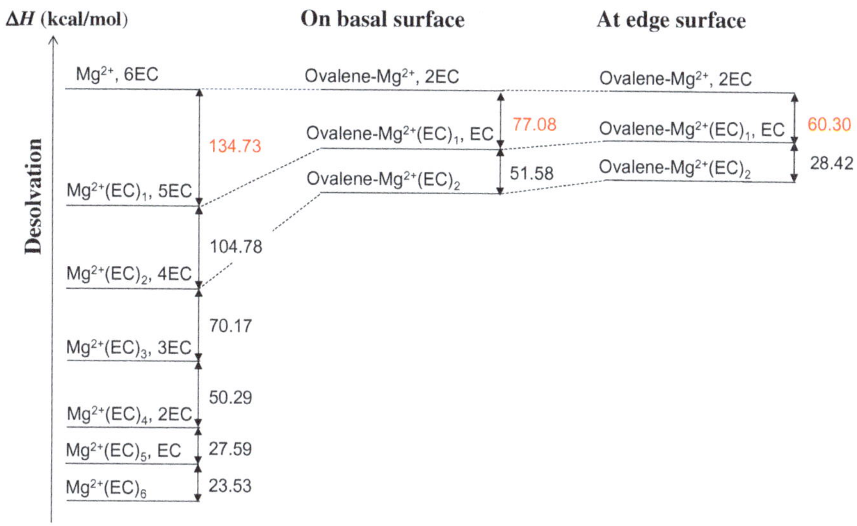

Fig. 4.25 Stepwise desolvation diagram of $Mg^{2+}(EC)_6$ with the enthalpy change (at 298.15 K and 1 atom) accompanied. The left represents the desolvation without the carbon electrode (see Fig. 4.23)

assumed to make solvation without dissociation into Li^+ and PF_6^-. In Fig. 4.26, the structures of solvated $LiPF_6$ with the solvent molecules EC or PC are shown, where they are supposed to be one-electron reduced upon the contact to the carbon negative electrode. This electron for the reduction can be accommodated either on a solvent molecule or on PF_6^- moiety, which is rather competitive. The reduced PF_6^- causes cleavage of the P–F bond to form LiF. In other words, when LiF coexists the solvent molecules do not solvate explicitly.

It might be of interest to compare the stabilization by the solvation between the events of one more solvent molecule and one more PF_6^- added to the solvated $LiPF_6$ sheath. In Fig. 4.27, EC and PC cases are shown. It is seen that in both solvents addition of one more PF_6^- generally causes larger stabilization than that of one more solvent molecule, and that stabilization difference ΔE is smaller in PC solution than in EC. That is, in EC, the stabilization by solvation is relatively smaller than in PC. This suggests that desolvation takes place more easily in $LiPF_6(EC)_4$ than in $LiPF_6(PC)_4$, which signifies a larger possibility of generation of the material by decomposing of EC than that of PC. The former decomposition product is mainly lithium ethylene dicarbonate (LEDC) and the latter lithium propylene dicarbonate (LPDC).

The optimized structures and the HOMO-LUMO gaps of these decomposition materials associated with two Li^+s are shown in Fig. 4.28 with those of other Li salts (Li_2CO_3 and LiF). The data of Li_2CO_3 is added for the reference. Note that the optimized structure of LPDC is rather "bent" being unfavorable for a compact stacking on the carbon negative electrode compared with LEDC and LiF. An isomer of LPDC, i.e., LPDC-L having a flat structure is also shown but this species is less stable than LPDC by 0.4 kJ/mol and the isomerization process from LPDC is

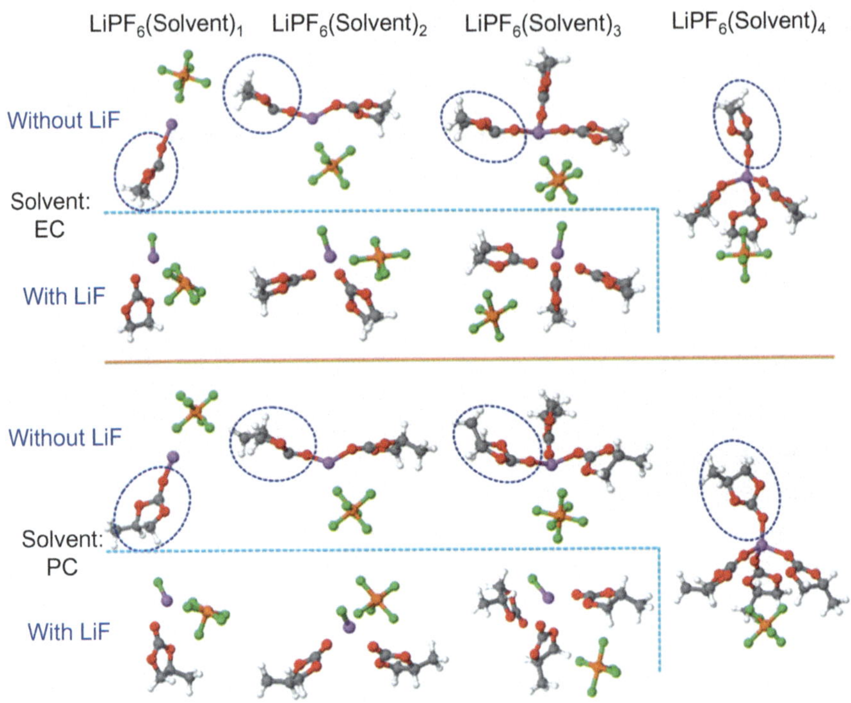

Fig. 4.26 The optimized structures of $LiPF_6(solvent)_n$ complexes ($n = 1–4$) after single-electron reduction. The reduced solvent molecule is shown by the dotted circle. When PF_6^- accepts the reducing electron, LiF is formed near the system. Reprinted with permission from Xing et al. (2018). Copyright 2018 American Chemical Society

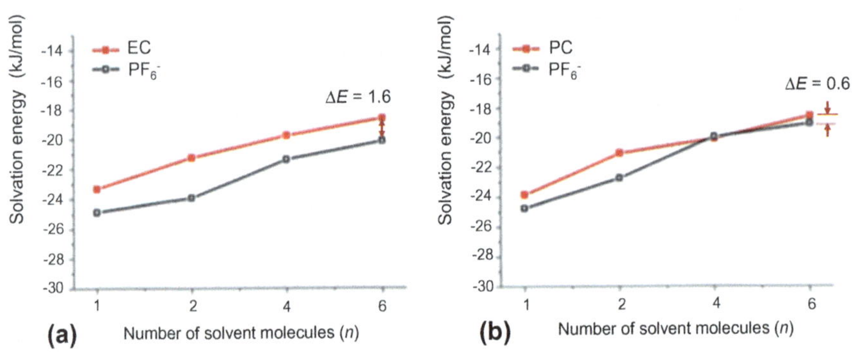

Fig. 4.27 Plots of {solvation energy of the n th solvent} defined by {$LiPF_6(solvent)_n$ – $LiPF_6(solvent)_{n-1}$ – solvent} (in red) and {solvation energy of PF_6^- with $Li^+(solvent)_n$ } defined by {$LiPF_6(solvent)_n$ – $Li^+(solvent)_n$ – PF_6^-} (in black). For **a** EC and **b** PC cases. Reprinted with permission from Xing et al. (2018). Copyright 2018 American Chemical Society

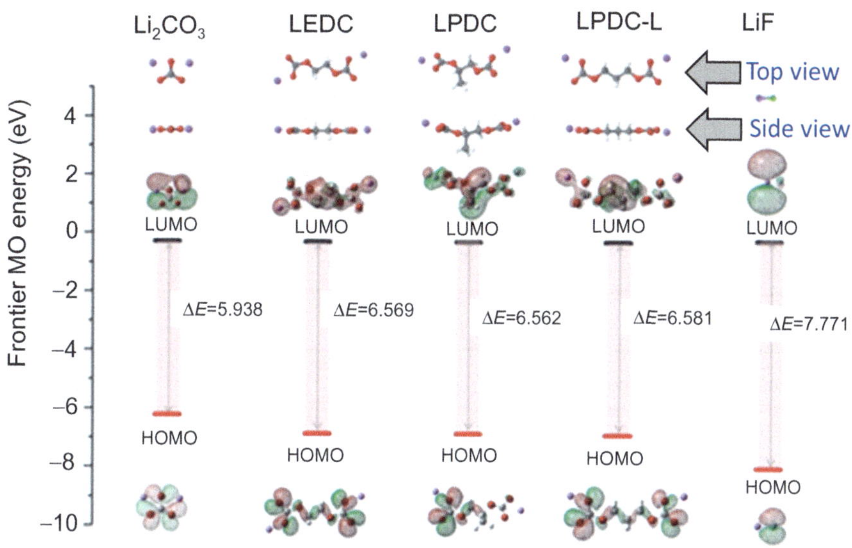

Fig. 4.28 Frontier MO patterns and their energy levels with the HOMO-LUMO gaps ΔE of lithium salts and $(Li^+)_2$(decomposed solvent) complexes. Their optimized structures are also shown on the top. Reprinted with permission from Xing et al. (2018). Copyright 2018 American Chemical Society

accompanied with large energy barrier (181.7 kJ/mol). The HOMO-LUMO gap can be index for the electronic insulation capability which is crucial to block the escape of electrons out of the negative electrode. The order of this tendency is LiF > LEDC $\simeq$ LPDC > Li_2CO_3 based on the energy gap. From all of these results and consideration of the molecular shape it can be conjectured that EC would make a better SEI with LiF compared with PC.

4.2.3 Molecular Design Toward Positive-Electrode Materials

In this subsection, an attempt to perform molecular design of novel positive electrode material for rechargeable batteries is introduced. Such materials should be able to construct multi-electron redox system, since that behavior is indispensable to the charging-discharging process. A series of derivatives of tetrathiafulvalene (TTF) molecule could be suitable for active materials for rechargeable batteries, since these derivatives have turned out to provide important roles acting as donors in a large number of molecular conductors and to show several one-electron redox waves (Yamada and Sugimoto 2004). A simple but crucial demand for the purpose described above is that the active materials in the electrode should have low solubility in the organic solvent used in the electrolyte solution. Toward achievement of this condition, the utilization of polymerized materials will be of use.

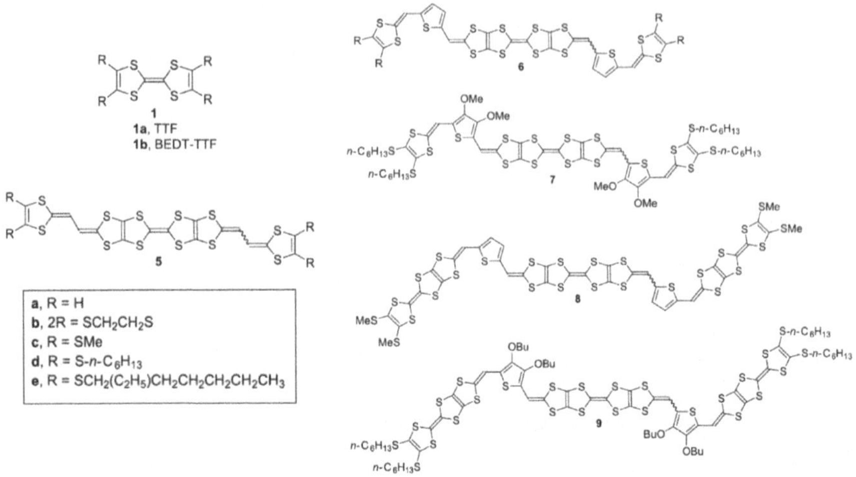

Fig. 4.29 Molecular structures of TTF and its derivatives. Adapted from Iwamoto et al. (2015) by The Authors licensed under CC BY 4.0 (https://creativecommons.org/licenses/by-nc-nd/4.0/deed.en)

Hence, as attempt to examine the possibility mentioned above, a series of extended version of tetrathiafulvalene (TTF) molecule have been considered as the candidates and actually synthesized based on the theoretical examination of their redox properties (Iwamoto et al. 2015). In Fig. 4.29are shown the molecular structures of TTF (**1**) , synthesized derivatives of vinyl extended tris-fused TTF molecule (**5**), and tris- and pentakis-fused TTF analogues extended by insertion of two thiophene rings (**6**–**9**). Each TTF fragment in these molecules can supply two electrons at maximum, so that in molecules **5**, **6**, and **7**, e.g., can become hexacation at maximum by stepwise oxidation. The multi-redox mode of these molecules (derivatives of **5** and **7**) is depicted in Fig. 4.30. The MO patterns concerning the multi-oxidation can be visualized. For instance, patterns and energies of several MO's near the HOMO of **5a** calculated by the DFT/B3LYP/6-31G* are shown in Fig. 4.31. It is seen that the HOMO pattern is distributed over the whole molecule, whereas the HOMO-1 pattern rather on the bilateral vinylogous TTF moieties. Moreover, the HOMO-2 is rather concentrated on the central TTF area as a whole. These MO's are all of π-type and are expected to guarantee rather robust molecular structures upon multi-redox cycles typically encountered in rechargeable batteries. This situation is almost the same with the MO patterns of **7a** though they are omitted here. The MO energies calculated for a couple of simplified molecules are listed in Table 4.3. These values are not much separated and will provide rather propitious oxidation behaviors upon electrochemical processes.

The redox behaviors of some molecules measured by cyclic voltammetry are shown in Fig. 4.32, which actually show multi-redox waves. There can be considered one-electron and two-electron redox waves in these considering the peak currents. The assigned electron numbers concerning the oxidation processes are also added at

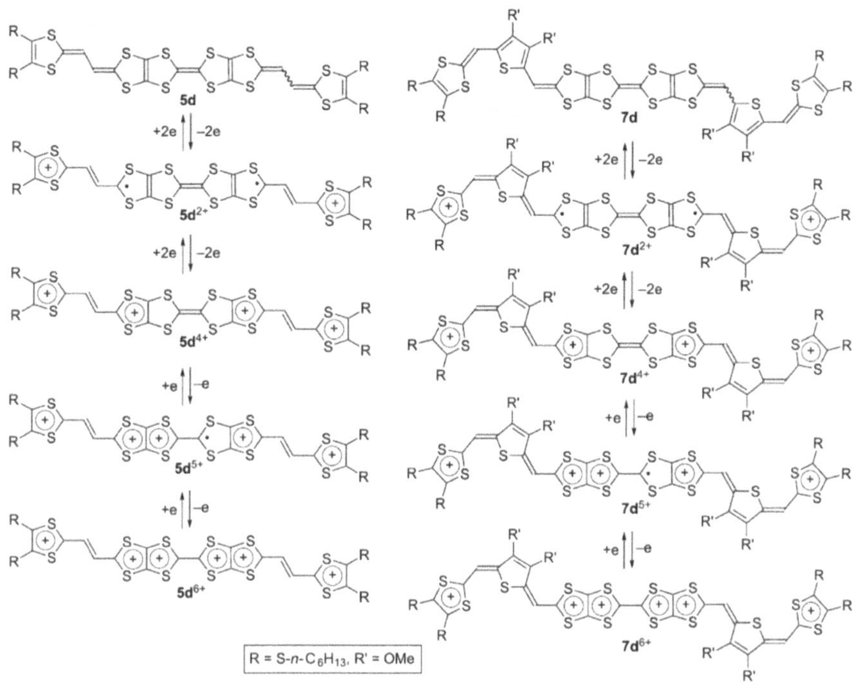

Fig. 4.30 Plausible redox processes of **5d** and **7d**. Adapted from Iwamoto et al. (2015) by The Authors licensed under CC BY 4.0 (https://creativecommons.org/licenses/by-nc-nd/4.0/deed.en)

the waves. Thus, **5d** and **7d** show six-electron oxidations, whereas **9d** ten-electron ones in accordance with the number of TTF motifs as described in the above. In **5d** and **7d,** the two positive charges formed by the first two-electron oxidation process can be distributed on rather whole molecule based on the HOMO pattern of **5d** in Fig. 4.31. Hence, two positive charges can possibly be separated over the whole molecular structure of **5d**$^{2+}$ and **7d**$^{2+}$ as much as possible in order to reduce the on-site Coulomb repulsion utilizing the delocalized HOMO pattern. Moreover, the two additional positive charges in **5d**$^{4+}$ and **7d**$^{4+}$ will be separately located each other on the two outer extended parts. However, in **5d**$^{6+}$ and **7d**$^{6+}$, the fifth and the sixth positive charges should cause larger on-site Coulomb repulsion since the HOMO-2 pattern has larger component in rather central part. For **9d**, similar analysis is also possible based on the MO patterns.

A coin-type battery was actually fabricated employing the positive electrode using **5b**, **5c**, **6b**, or **8c** as the active material with acetylene black as the conductive additive and poly(tetrafluoroethylene) as the binder. The electrolyte solution was prepared by mixing ethylene carbonate (EC) and diethyl carbonate (DEC) (1:5, v/v) with 1.0 M $LiBF_4$, and Li was employed as the negative electrode. The cell showed initial discharge capacities of 157–190 mAh/g and initial energy densities of 535–680 mWh/g. The initial discharge capacities became 64–86% after 40 cycles. The

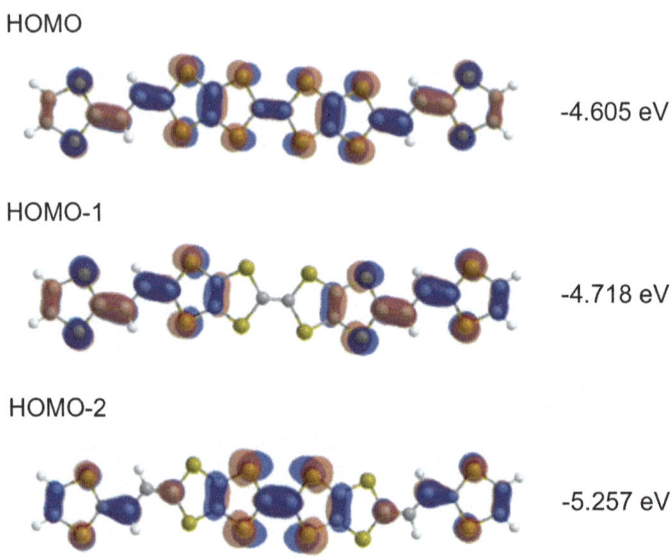

Fig. 4.31 Patterns and energies of the HOMO and the subjacent MO's of **5a**. Adapted from Iwamoto et al. (2015) by The Authors licensed under CC BY 4.0 (https://creativecommons.org/licenses/by-nc-nd/4.0/deed.en)

Table 4.3 Several π-type MO energies[1] of 5a, 6a, and 8a in Fig. 4.29[2]

	5a	6a	8a
HOMO	−4.605	−4.532	−4.602
HOMO-1	−4.718	−4.589	−4.643
HOMO-2	−5.257	−5.129	−4.967
HOMO-3	–	–	−5.061
HOMO-4	–	–	−5.328

Adapted from Iwamoto et al. (2015) by The Authors licensed under CC BY 4.0 (https://creativecommons.org/licenses/by-nc-nd/4.0/deed.en)

[1]In eV

[2]The calculations including the structure optimization were performed by DFT/B3LYP/6-31G*

initial discharge capacities thus observed are comparable with or more than that of commercially available lithium ion batteries (LIBs) on the market (150–170 mAh/g). The initial energy densities are also superior to those obtained in most of inorganic positive electrode materials used in LIBs (Aravindan et al. 2013; Thackeray et al. 2012).

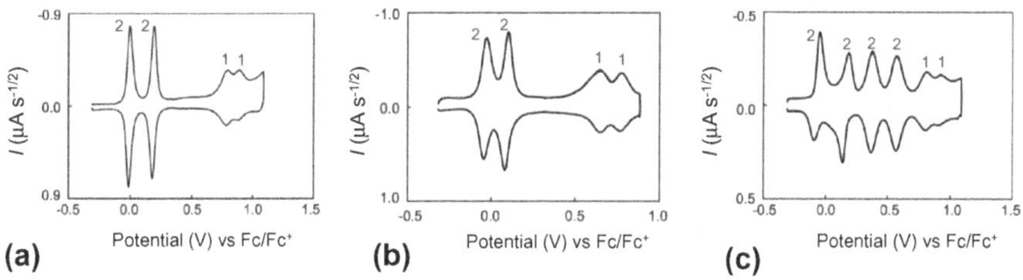

Fig. 4.32 Deconvoluted cyclic voltammograms of **a 5d**, **b 7d**, and **c 9d** in CS_2/benzonitrile (1:1, v/v) solution. Numbers near each oxidation peak signify the electron numbers concerned. Adapted from Iwamoto et al. (2015) by The Authors licensed under CC BY 4.0 (https://creativecommons.org/licenses/by-nc-nd/4.0/deed.en)

4.3 Catalytic Chemistry

Catalytic action is one of the most important phenomena involved in chemical reaction and understanding of this action is of keen interest in theoretical and computational chemistry. It is rather recent years for computational chemistry to be able to perform quantitative calculations for catalytic reactions due to development of computation ability (Frenking 2005). The calculations concerning catalytic action have mostly been devoted to structural change of chemical species in the catalytic cycle and energetic estimation connected to the reaction barrier (activation energy) and the energy difference between the initial and the final states during the reaction to clarify whether that is exothermic or endothermic.

In addition to these, important mission of theoretical chemistry would consist of decomposition of catalytic-reaction cycle into elementary reactions and their mechanisms and clarifying their roles in the overall reaction. Based on such information understanding of the total feature of the chemical reaction including catalytic cycle would become possible. In this section, examples of attempts to clarify catalytic action by theoretical chemistry are to be afforded.

4.3.1 σ-Bond Activation by Organometallic Complex Catalyst

Organometallic complexes of transition metals such as Ru, Rh, Pt, and Pd are of importance since they are used as homogeneous catalysts available for many useful chemical reactions including coupling reactions toward bond formation of C–C, C–N, and so on. For instance, Pd or Pt complex plays important role as the catalyst in the cross-coupling reaction

$$\text{R1-X} + \text{R2-Y} \rightarrow \text{R1-R2} \quad (4.3)$$

R1 and R2 signifying, e.g., hydrogen atom, alkyl, vinyl, phenyl, or aryl group and X and Y anionic species. When R1 = R2 and X = Y, it is particularly called homo-coupling reaction. A typical example of Eq. (4.3) is Suzuki-Miyaura cross-coupling reaction (Miyaura et al. 1979) expressed as

$$\mathrm{R1\text{-}Br} + \mathrm{R2\text{-}B(OH)_2} \underset{\mathrm{Base}}{\overset{\mathrm{Pd(PPh_3)_4}}{\rightarrow}} \mathrm{R1\text{-}R2} \tag{4.4}$$

in which Pd(0) complex participates as the catalyst. A possible reaction mechanism using Pd(0) has been proposed by the diagram in Fig. 4.33. Another type of coupling reaction has also been developed to use Pd(II) complex like $PdCl_2$ (Moritani and Fujiwara 1967) as in

$$\mathrm{Ar\text{-}H} + \mathrm{Ar(CH=CH)\text{-}H} \underset{\mathrm{Benzene+AcOH}}{\overset{\mathrm{PdCl_2}}{\rightarrow}} \mathrm{Ar\text{-}(CH=CH)\text{-}Ar} \tag{4.5}$$

In the both of these reactions, the initial σ-bond activation is one of the most fundamental and crucial processes as an elementary reaction, since this is the essential step to introduce new functional groups via cleavage of the σ-bond and formation of C–M bond (M; transition metal). There can be considered two major kinds of the cleavage patterns, i.e.,

(i) $\mathrm{R1\text{-}R2} + \mathrm{ML_n} \rightarrow \mathrm{M(R1)(R2)L_n}$ Oxidative addition
(ii) $\mathrm{R1\text{-}R2} + \mathrm{MXL_n} \rightarrow \mathrm{M(R1)L_n} + \mathrm{R2\text{-}X}$ Heterolytic (ionic) activation

where L designates ligand species. In the oxidative addition (i), R1 and R2 are considered to be both anionic when they attach M after the σ-bond cleavage and, hence, M becomes oxidized. On the other hand, in the heterolytic activation (ii), R1 and R2 become anionic and cationic, respectively, after the cleavage. It is noted that

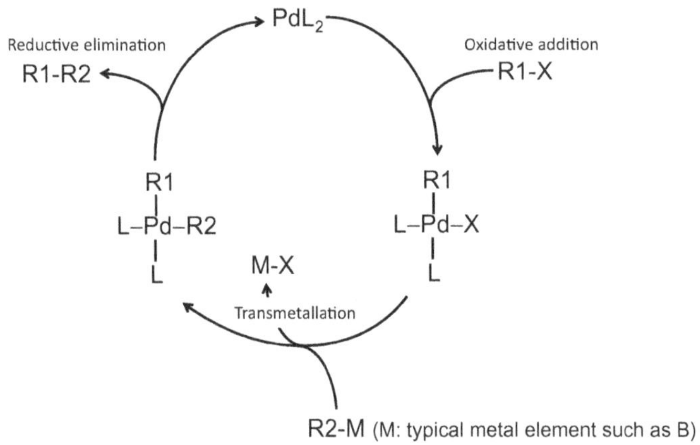

Fig. 4.33 A possible reaction mechanism for cross-coupling reaction using Pd(0) catalyst. From Sakaki et al. (2010), Copyright © 2010 by John Wiley & Sons, Inc

in (i) the metal oxidation state increases by 2, and that in (ii) the metal oxidation state does not change.

In the following are to be mentioned theoretical analyses with respect to catalytic behavior of organometallic complexes (Biswas et al. 2000; Sakaki et al. 2010). All the molecular geometries including the transition states were obtained by the calculation based on DFT/B3LYP/Basis set I (see below). Table 4.4 lists the calculation results of energies as to typical elementary reactions for C–H σ-bond activations of methane and benzene molecules calculated by MP4(SDQ)/Basis sets II and III (see below). As the Basis set I the effective core potentials (ECPs) were employed for core electrons of Pd (up to 3d), Pt (up to 4f), and P (up to 2p). For valence electrons of Pd, Pt, and P were used (311/311/31), (311/311/111), and (21/21/1), respectively. For C and H were used MIDI-3 and (31), respectively. A p-polarization function ($\zeta = 1.0$) was added to the active H liberating from the C–H bond. For O, a (421/211) set with p-diffuse function ($\zeta = 0.059$) was used. As to details in Basis sets II and III readers are encouraged to see the original reference (Biswas et al. 2000).

The calculated data suggest that for C–H σ-bond activation of methane the Pt(II) catalyst is the most favorable due to the smallest activation energy and the largest exothermicity compared with Pd(0) and Pd(II), which is in agreement with the experimental data for the conversion reaction from methane to methanol (Periana et al. 1998). On the other hand, for C–H σ-bond activation of benzene has been shown that Pd(II) catalyst is the most favorable compared with Pt(II) and Pd(0) due to the smallest activation energy.

It would be of useful to examine the origin of the magnitude of reaction energy for the C–H σ-bond activation of benzene in Pd(II) catalyst, i.e., $Pd(O_2CH)_2$ in this calculation compared with that in Pd(0) catalyst, $Pd(PH_3)_2$. For the Pd(II) there essentially occurs the formation of O–H bond (underlined in the reaction below) by the attach of H^+ to O_2CH^- as seen in the reaction

Table 4.4 Energy data[1] for C–H σ-bond activation of methane and benzene by Pd and Pt catalysts (in kcal/mol)[2]

Oxidation state of metal	Methane		Benzene	
	E_a	ΔE	E_a	ΔE
Pd(II)[3]	21.5	−8.3	15.7	−17.2
Pt(II)[4]	17.3	−13.3	20.9	−24.1
Pd(0)[5]	34.7	31.5	26.5	22.1

[1] Activation energy E_a and energy difference ΔE (the difference between the product and the summation of reactant). Negative ΔE signifies the exothermic and positive endothermic reactions

[2] Calculated by MP4(SDQ)/Basis sets II and III. See ref. in the above for details of the calculation basis set

[3] $Pd(O_2CH)_2$

[4] $Pt(O_2CH)_2$

[5] $Pd(PH_3)_2$

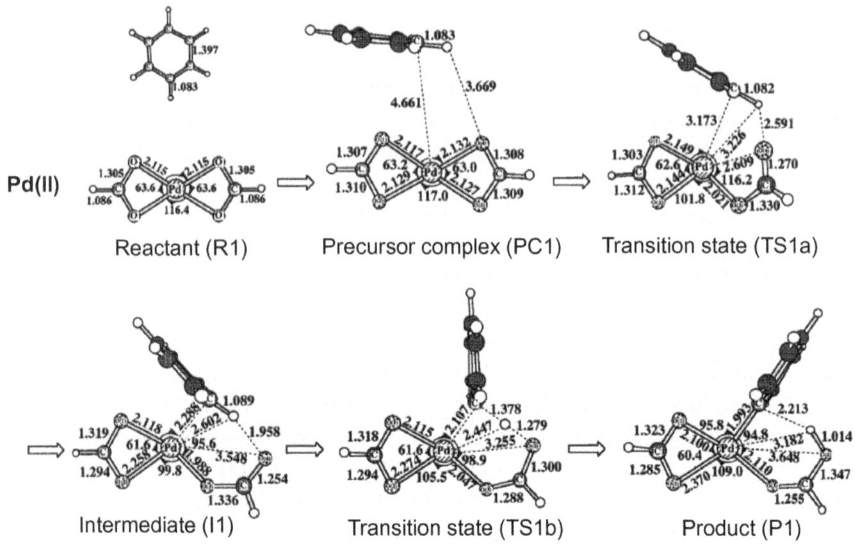

Fig. 4.34 Structural changes in the C–H σ-bond activation of benzene by $Pd(O_2CH)_2$. Bond lengths and angles are shown in Å and degrees, respectively. Reprinted with permission from Biswas et al. (2000). Copyright 2000 American Chemical Society

$$Pd(O_2CH)_2 + C_6H_6 \rightarrow Pd(C_6H_5)(O_2CH)(HCOOH) \tag{4.6}$$

in which the O–H bond formation energy was obtained as 114.4 kcal/mol. This value is large enough to cause exothermic energy as seen in Table 4.4. The calculated structural changes throughout the reaction indicated by Eq. (4.6) are shown in Fig. 4.34. There are seen two transition states TS1a and TS1b. In TS1b, the C–H bond is almost broken and the Pd–C bond is getting to be shorter. It can be said in TS1b that O–H bond formation is still proceeding but Pd–C bond is almost formed. The change in natural bond orbital (NBO) populations of the concerning atoms and group along this reaction is plotted in Fig. 4.35. The electron population changes of the active H and the remaining C_6H_5 rather indicate the occurrence of polarization $(C_6H_5)^{\delta-}$–$H^{\delta+}$, showing tendency of heterolytic activation process. This fact will enhance the electron transfer from C_6H_5 to Pd(II) or Pt(II). In other words, Pd(II) or Pt(II) catalyst causes electrophilic attack to the benzene fragment.

On the other hand, for Pd(0) catalyst there occurs the formation of Pd–H bond as seen in

$$Pd(PH_3)_2 + C_6H_6 \rightarrow Pd(C_6H_5)(PH_3)_2(H) \tag{4.7}$$

whose formation energy was calculated to be 48.5 kcal/mol being less than half of the O–H bond. The structural changes and the change in NBO populations accompanied with this reaction are shown in Figs. 4.36 and 4.37, respectively. In Fig. 4.36, there are seen three kinds of precursor complex, each of which has different η numbers.

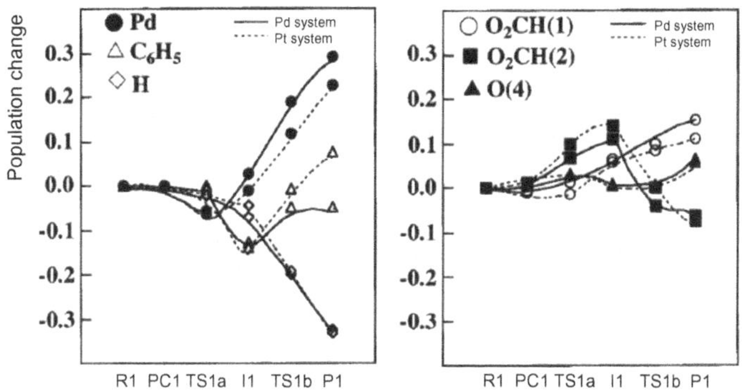

Fig. 4.35 Electron population changes in the C–H σ-bond activation of benzene by $Pd(O_2CH)_2$. Note that the increase of population change signifies that the atom or atomic group becomes more negative. Also shown for $Pt(O_2CH)_2$. See Fig. 4.34 with respect to the horizontal axis. $O_2CH_2(2)$ reacts with the active H of benzene. Reprinted with permission from Biswas et al. (2000). Copyright 2000 American Chemical Society

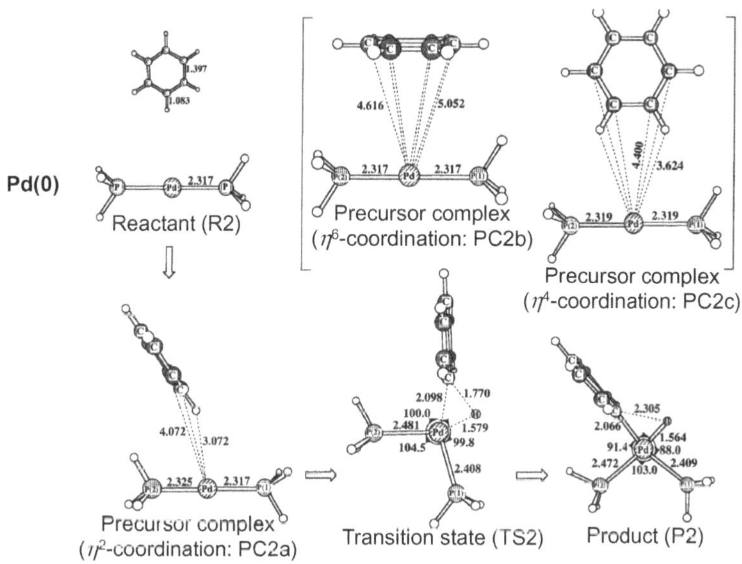

Fig. 4.36 Structural changes in the C–H σ-bond activation of benzene by $Pd(PH_3)_2$. Bond lengths and angles are shown in Å and degrees, respectively. Reprinted with permission from Biswas et al. (2000). Copyright 2000 American Chemical Society

It is also seen that in PC2a the Pd–C and Pd–H bonds are considerably long which is reminiscent of van der Waals complex unlike the precursor I1 in Fig. 4.34. The population changes of both the activated H and the remaining C_6H_5 become more negative as shown in Fig. 4.37 according to the reaction process, whereas those of Pd

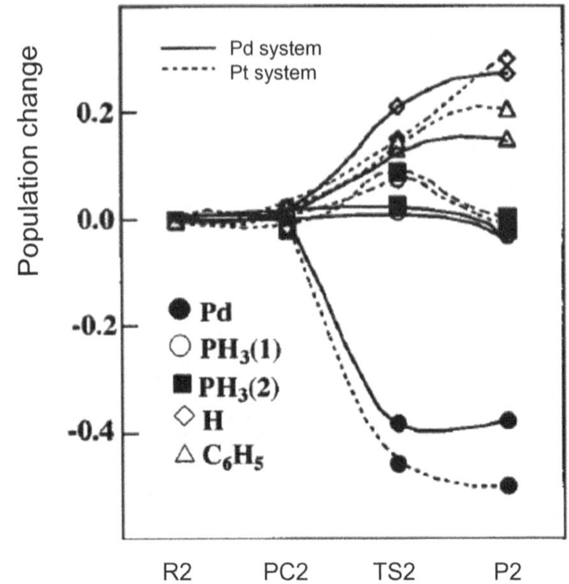

Fig. 4.37 Electron population changes in the C–H σ-bond activation of benzene by $Pd(PH_3)_2$. Note that the increase of population change signifies that the atom or atomic group becomes more negative. Also shown for $Pt(PH_3)_2$. See Fig. 4.36 with respect to the horizontal axis. Reprinted with permission from Biswas et al. (2000). Copyright 2000 American Chemical Society

and Pt more positive. This feature implies that the original Pd(0) and Pt(0) become oxidized, being characterized by an oxidative addition, which is completely different from what is found in Fig. 4.35.

Thus, the heterolytic cleavage to yield H^+ and $C_6H_5^-$ by Pd(II) becomes dominant due to large stabilization from O–H bond formation compared with oxidative addition in case of Pd(0).

4.3.2 *Reaction Design Toward Photoreduction of CO_2*

Photoreduction processes of CO_2 would be one of the key technologies to resolve the global warming issues and have been being kept as challenging theme to many researchers in chemistry. Organometallic complexes could form useful photocatalytic system for this purpose and several metallic species such as Ru (Lehn and Ziessel 1990; Nazeeruddin et al. 1993; Kuramochi et al. 2014), Re (Takeda et al. 2008), Ru–Re (Koike et al. 2009) binuclear system, and so on have been examined. In this connection, a computational analysis toward fabrication of a photoreduction system of CO_2 is to be introduced in the following by employing a Ru complex, $RuL_2(NCS)_2$ (L = 2,2′-bipyridyl-4,4′-dicarboxylic acid) denoted as N3 (Fueno et al. 2015). The calculation was done by DFT/B3LYP method in which SDD (Stuttgart/Dresden effective core potential) was used for Ru metal and 6-31G** for other elements.

The N3 complex shown in Fig. 4.38a or its analogues are the dyes utilized as photosensitizing material for, e.g., TiO_2 electrode of dye sensitized solar cell. Calculated

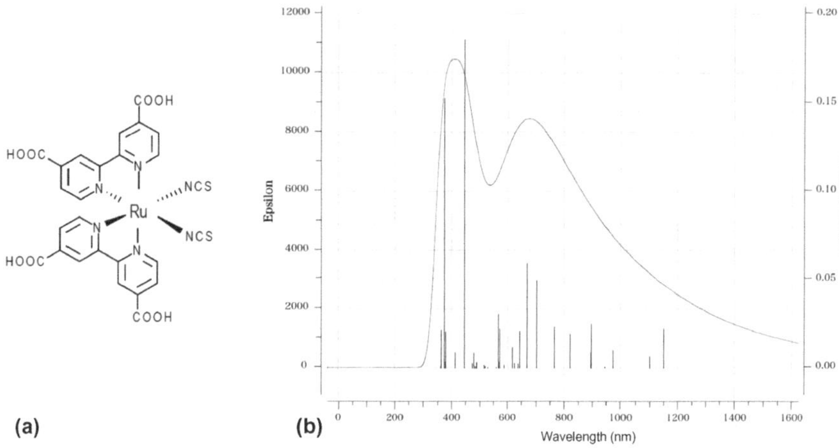

Fig. 4.38 **a** Molecular structure of N3 and **b** the calculated optical spectrum

optical absorption energies of N3 are listed in Table 4.5 which is in reasonable agreement with the experimental data (Nazeeruddin et al. 1993). The calculated optical absorption spectrum and the corresponding MO's are shown in Figs. 4.38b and 4.39, respectively. It is understood that this complex has favorable spectrum being available up to ca. 700 nm or more. It is seen that the calculated peak at 445 nm is related to the transition from the HOMO-6 to the LUMO signifying the electron transfer from the metal (Ru) area to the ligands, which corresponds to, what is called, the metal-ligand charge transfer (MLCT). This phenomenon is first related to generation of the ^{1}MLCT, which rapidly changes into the ^{3}MLCT state through spin-orbit coupling due to heavy atom effect of Ru. As seen in Table 4.6, the natural charge of Ru at ^{3}MLCT of **a–1** increases by ca. 0.2 actually signifying the occurrence of the

Table 4.5 Major peaks of the calculated optical absorption spectrum of N3

No of excitation	Wavelength of the peak (in nm)	Corresponding MO transition in TD-DFT	Coefficient of the MO transition	Oscillator strength
10th	668.76	HOMO → LUMO + 3	0.6408	0.0591
9th	704.08	HOMO → LUMO + 2	0.6575	0.0496
29th	445.12	HOMO-6 → LUMO	0.5866	0.1853
33rd	373.66	HOMO-5 → LUMO + 3	0.5408	0.1523
		HOMO-4 → LUMO + 2	0.4082	

[1]Experimental absorption peaks are at 534 nm (1.42), 396 nm (1.40), and 313 nm (3.12), where values in parentheses indicate ε in 10^4 /M cm (Nazeeruddin et al. 1993)

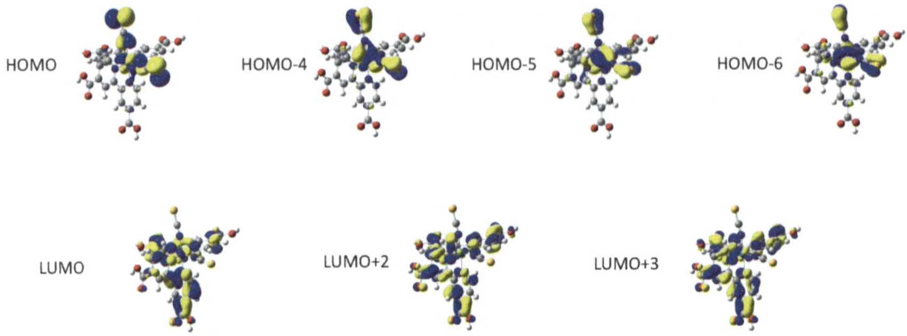

Fig. 4.39 MO patterns of N3 responsible for the optical absorption listed in Table 4.5

Table 4.6 Calculated data at each state (see Fig. 4.41)

State	**a–1**	**a–1** ^{3}MLCT	**a–2**	**a–3**	**a–4**
Ru–N bond length[1] (in Å)	2.043	2.001	2.062	2.030	1.995
Natural charge of Ru	0.5483	0.7104	0.5638	0.6531	0.6755
Relative energy[2] (in eV)	0 (0)	+0.966 (+0.576)	−2.297 (−2.254)	+0.047 (+0.123)	+5.006 (+4.839)

[1]Between Ru and N atoms of Ru–NCS
[2]In parentheses are those for N3-polyene system

charge transfer from Ru to ligands. It is noted that in **a–2** being anionic the natural charge on Ru is almost the same with that in **a–1**, signifying excessive electron in **a–2** ought to be on the ligands.

The total catalytic cycle of photoreduction of CO_2 has been proposed as starting from the **a–1**, ^{3}MLCT of **a–1**, **a–2**, **a–3**, and then to the CO_2 adduct liberating HCOOH or CO followed by changing to **a–4** and finally to the **a–1** again as shown in Fig. 4.40. The optimized structures of all these states are shown in Fig. 4.41. The virtue of utilization of the ^{3}MLCT of **a–1** can be understood by that electron is easily injectable to the lower SOMO eventually yielding an anionic species as illustrated in Fig. 4.42.

There can be considered two ways after formation of the CO_2 adduct to the **a–3** state as shown in Fig. 4.43 in the form of O=C=O or COO adduction. From the former adduct, $HCOO^-$ is liberated and from the latter CO. The energy diagrams of these two ways are indicated in Figs. 4.44 and 4.45, respectively.

It might be possible to reinforce the optical absorption of N3 by introducing a polyene (dodecaene) substructure. An example of molecular design of N3-polyene is shown in Fig. 4.46a. It is seen from the calculated optical absorption spectrum in Fig. 4.46b the tail appreciably extends to the near IR region, which ensures the utilization of solar light becomes more effective. Moreover, from Table 4.7, that the oscillator strength of the absorption in this system becomes much larger compared

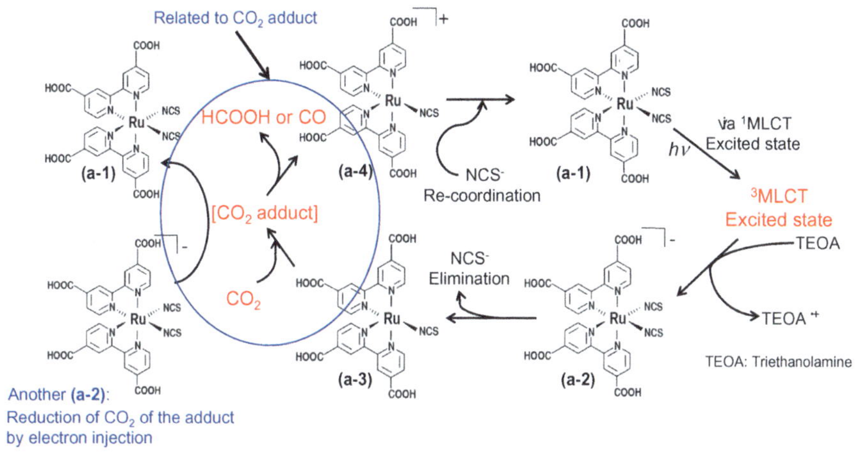

Fig. 4.40 A proposed catalytic cycle of CO_2 reduction utilizing the Ru-complex N3. MLCT signifies metal -ligand charge transfer

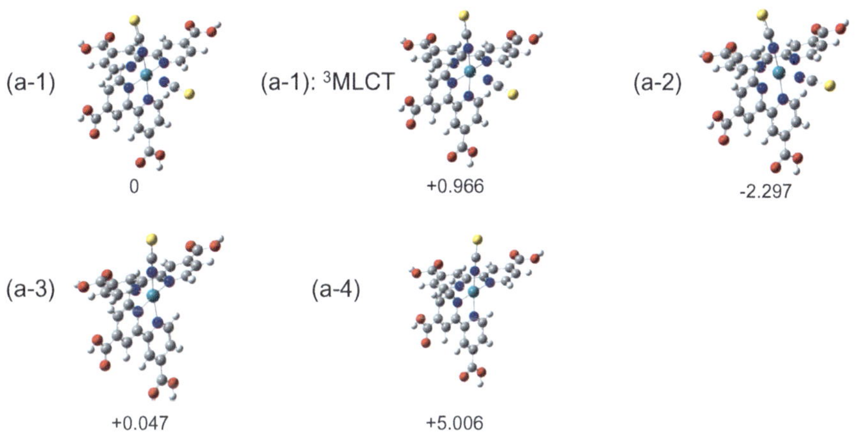

Fig. 4.41 Optimized structures of N3 at each state with the relative energy (in eV). Note that energies of (**a–3**) and (**a–4**) show the summation with that of the eliminated NCS^-

with those of N3. The MO patterns corresponding to the optical absorption in Fig. 4.47 shows the contribution of optical transition from the main chain of polyene to the N3 moiety in addition to the transition within the polyene chain.

This enhancement of the oscillator strength f comes from that of the transition moment $\mathbf{m}_{ij}$ between the MO levels i and j (such as the HOMO and the LUMO) described as follows:

$$f = 1.085 \times 10^{-5} \Delta E_{ij} \cdot \mathbf{m}_{ij}^2 \qquad (4.8)$$

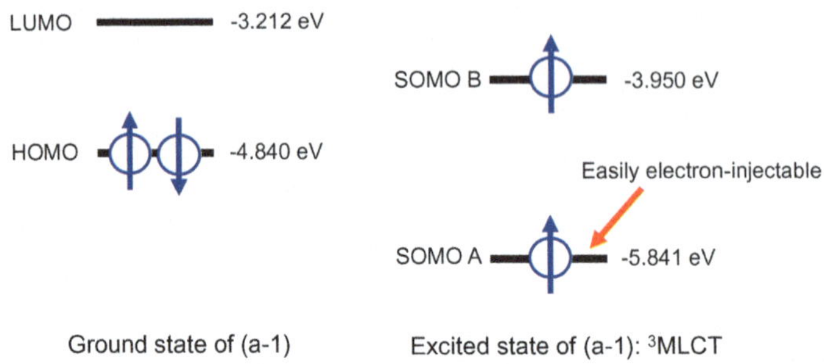

Fig. 4.42 Schematic energy diagrams of N3 working as the reductive catalyst taking advantage of ^{3}MLCT state. MLCT signifies metal to ligand charge transfer

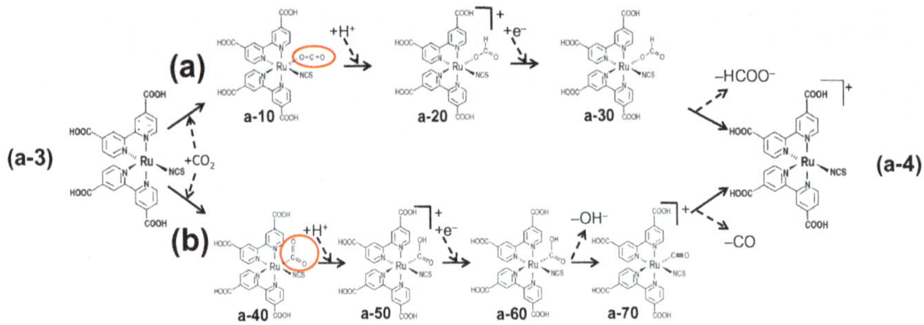

Fig. 4.43 Two kinds of CO_2 reduction pathways via **a** OCO and **b** COO additions to (**a–3**) state with various intermediates

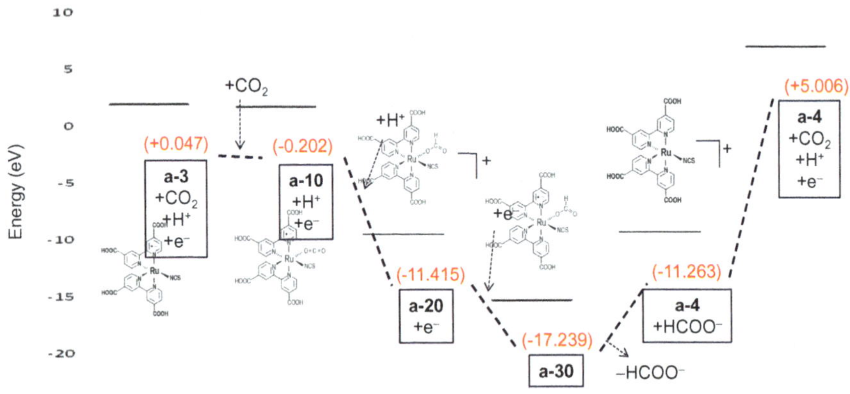

Fig. 4.44 A series of relative energy changes in the Ru–OCO cycle. See Fig. 4.41(**a–1**) as to setting of the zero energy

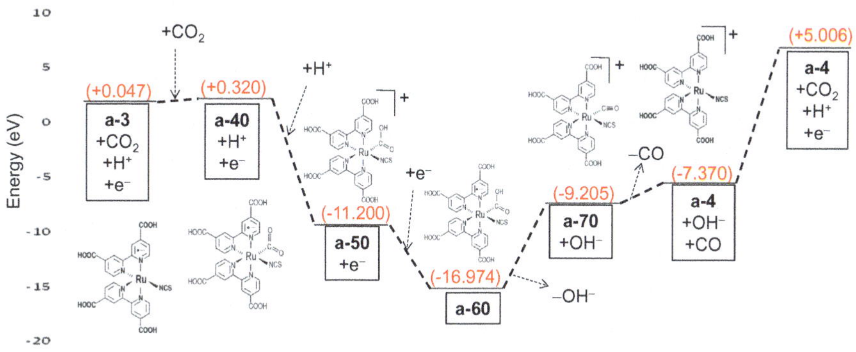

Fig. 4.45 A series of relative energy changes in the Ru-COO cycle. See Fig. 4.41(**a–1**) as to setting of the zero energy

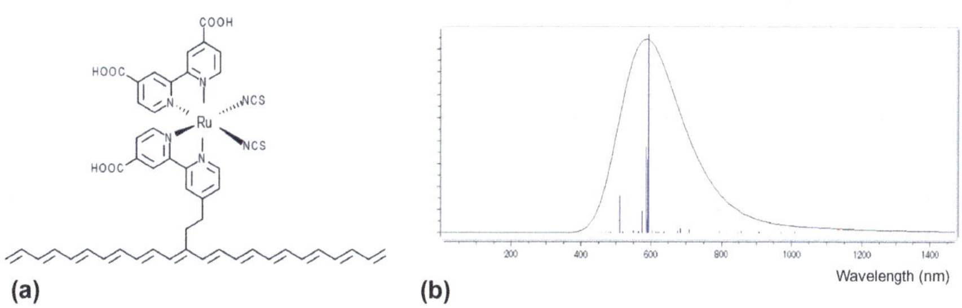

Fig. 4.46 **a** Molecular structure of N3-polyene and **b** the calculated optical spectrum

Table 4.7 Major peaks of the calculated optical absorption spectrum of N3-polyene

No of excitation	Wavelength of the peak(in nm)	Corresponding MO transition in TD-DFT	Coefficient of the MO transition	Oscillator strength
19th	598.91	HOMO → LUMO + 2	0.5157	1.9269
		HOMO-5 → LUMO	0.4018	
20th	595.71	HOMO → LUMO + 2	−0.3085	0.7083
		HOMO-5 → LUMO	0.5433	
22nd	592.69	HOMO-1 → LUMO + 5	0.5176	0.8305

$$\left(\mathbf{m}_{ij}\right)_x = \sqrt{2}\left(\sum_{r}^{\text{all AO's}} c_r^i c_r^j \bar{x}_{rr} + 2\sum_{r>s}^{\text{all AO's}} c_r^i c_s^j \bar{x}_{rs}\right) \tag{4.9}$$

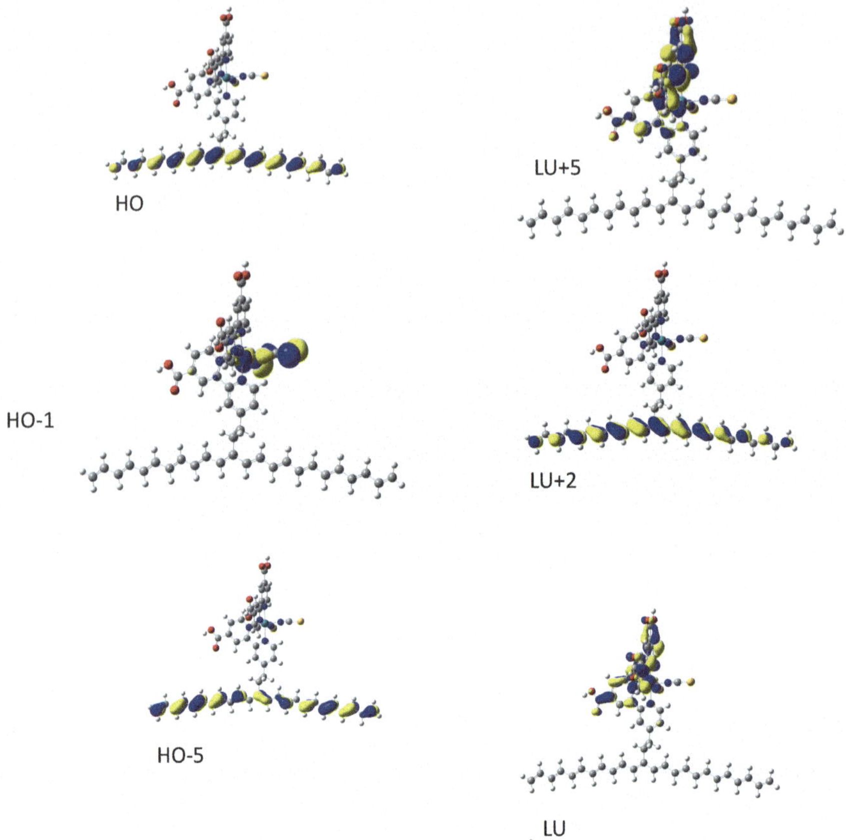

Fig. 4.47 MO patterns of N3-polyene concerning the optical absorption listed in Table 4.7

where the off-diagonal elements in the second term of the right-hand side of Eq. (4.9) cross the polymer part and N3. Here, ΔE_{ij}, c_r^i, and $\bar{x}_{rs}$ stands for the energy difference of the ith and jth MO levels expressed in cm^{-1} unit, the ith MO coefficient for the rth AO χ_r, and expectation value of x averaged by AO's χ_r and χ_s, respectively. Thus, the existence of polyene as a kind of catalytic support for N3 is appreciable to enhance the optical absorption. This occurs without disturbance of the whole catalytic cycle as shown in Table 4.6. Such catalytic support has also been attempted by adding C3N4 for Ru complex without clarification at molecular level (Kuriki et al. 2015).

4.4 Theoretical Biology

Biological systems of usual interest are usually too large in size to be dealt with by full quantum mechanics (QM) calculation apart from rather simplified case. This is obviously related to lack in computation-machine resource at present. Hence, analyses of biological systems by combination of quantum mechanics (QM) and molecular mechanics (MM) or by molecular dynamics (MD) ought to be employed, which is still one of the most challenging themes in the current theoretical chemistry. In this section, some recent challenges to these themes from the theoretical side are to be introduced.

4.4.1 QM/MM Study of Catalytic Action of Cytochrome P450cam

Cytochrome P450 (also called CYP) is one of the hemoproteins and belongs to monooxygenase family working as a redox enzyme. P450 widely distributes among biological kingdom including microbials, plants, and animals. Especially P450cam is one of P450 families and makes hydroxylation at 5 exo position of camphor. In the catalytic cycle of P450cam the active site is the heme iron in a protoporphyrin IX complex with a cysteine (Cys) residue as the axial ligand and a molecular oxygen first bound to yield a ferric hydroperoxo species (Compound 0 in Fig. 4.48). The conversion process from the Compound 0 to the active iron-oxo species 1 is so rapid that the mechanism therein had not been well understood experimentally. For this reaction there can be two possible pathways as illustrated in Fig. 4.48. That is, in Mechanism I was assumed that protonation of the oxygen first occurs and then the O–O bond cleavage takes place, whereas in Mechanism II occurs the contrary situation.

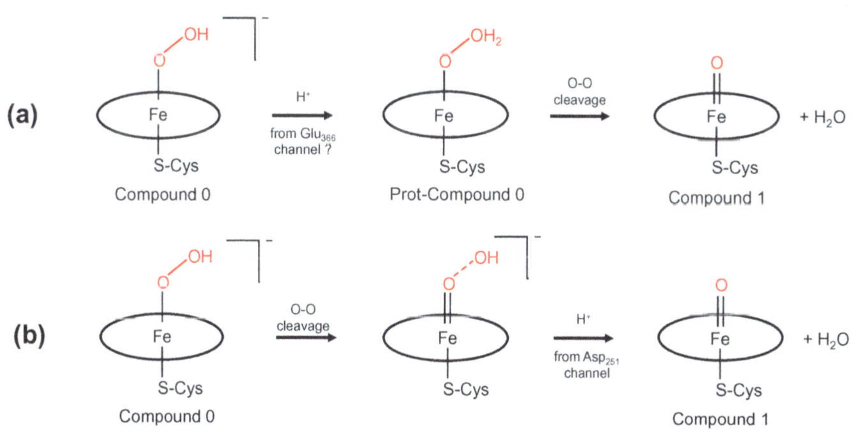

Fig. 4.48 Two kinds of mechanisms considered from a ferric hydroperoxo species of P450cam to its active iron-oxo ligand state: **a** Mechanism I and **b** II. Reprinted with permission from Zheng et al. (2006). Copyright 2006 American Chemical Society

The calculated reaction-energy profile based on a combined examination of quantum mechanical and molecular mechanical (QM/MM) methodologies with the DFT/UB3LYP/B1, B2, and B3 (see Sects. 3.2 and 3.6) for QM and the CHARMM22 force field for MM (see Sects. 3.4 and 3.6) implies that Mechanism II is the more favorable, in which the initial O–O bond cleavage is followed by a proton transfer with a simultaneous electron transfer to yield Compound 1 (Zheng et al. 2006). The calculation results have predicted that the energy barrier of O–O bond cleavage as the rate-determining step is about 13–14 kcal/mol in Mechanism II and that proton is plausibly supplied via Asp_{251} channel connected to the protein surface. On the other hand, all the reaction paths in Mechanism I requires large activation energies. This analysis is an example of clarification of enzyme reaction near its active site by the calculation.

4.4.2 MD Simulation of Viruses

Along with the recent explosive acceleration of capability of computational resources, the system size and the length for simulation time by the molecular dynamics (MD) calculation has been much enhanced. Recent target system by the MD analysis includes more than ten million atoms and to simulate the motion of the total system in μs order. This is essential for analyses of large-scale systems such as viruses in the biological field. The current computational situation has made it possible to treat such huge biological systems compared with the past by the MD calculations and two examples are to be given in what follows.

4.4.2.1 Satellite Tobacco Mosaic Virus (STMV)

The first all-atom molecular dynamics (MD) simulation of the satellite tobacco mosaic virus (STMV) shown in Fig. 4.49 has been carried out (Freddolino et al. 2006). The STMV dealt with consists of a small icosahedral capsid with diameter 16 nm composed of 60 proteins (each MW = 14,500) and an RNA with 1058 base pairs. The MD simulation is performed with the use of CHARMM22 force field for proteins, CHARMM27 for nucleic acids, and TIP3P for water molecules. The total number of atoms is about 1 million including water molecules as the solvent.

Simulation for the period of 13 ns has been performed to find that capsid is structurally stabilized by presence of RNA in accordance with the hypothesis based on the experimental evidence (Kuznetsov et al. 2005). In Fig. 4.50, the collapse behavior of capsid in the absence of RNA is shown. Moreover, it has been found out Mg^{2+} ions added for compensation of negative charge of RNA were attaching to the RNA during the simulation time.

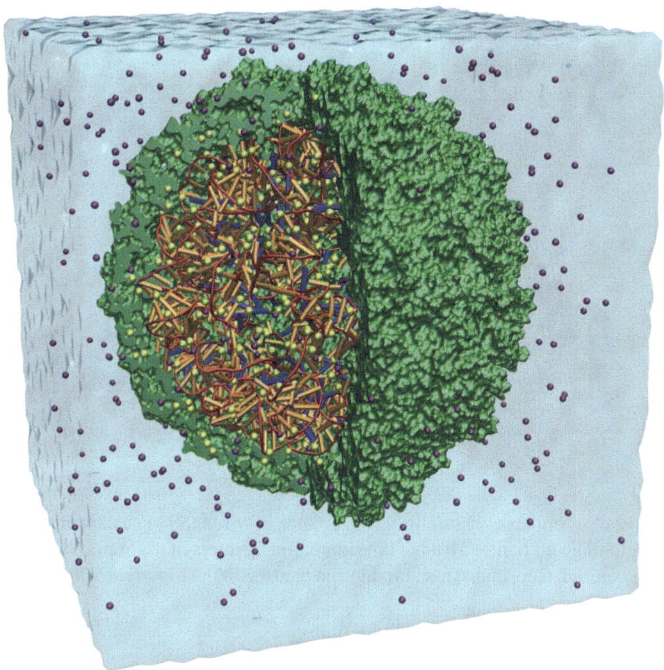

Fig. 4.49 Illustrative expression of a satellite tobacco mosaic virus (STMV) particle in a box of $220 \times 220 \times 220$ Å^3 with water as solvent molecules. The capsid is indicated by green color and the RNA inside is shown by cutting out. Reprinted from Freddolino et al. (2006).Copyright 2006, with permission from Elsevier

4.4.2.2 Poliovirus

More recently, all-atom MD calculation has been performed for empty poliovirus capsid (Andoh et al. 2014). This system consists of very large numbers of molecules (6.5 million) and the MD simulation has been achieved for sufficiently long time (200 ns). The K-computer at RIKEN in Japan (Yonezawa et al. 2011) was employed to perform the MD calculation. The CHARMM22 potential was applied for the proteins and electrolytes, TIP3P for water. After achievement of the initial equilibration process, the calculation has been performed in the NPT ensemble at $P =$ 0.1 MPa and $T = 310.15$ K. More details are to be referred to the original article.

The result of equilibrium state is shown in Fig. 4.51. The virus capsid has been shown to exchange water molecules with high speed between the inside and the outside thereof so as to result in achievement of equilibrium of hydrostatic pressure (Fig. 4.52). This behavior assures prompt equilibrium of hydrostatic pressure with exchanging water molecules under high pressure. However, within the time of the simulation the capsid does not exchange ions present in the solution indicating the behavior as semi-permeable membrane. Moreover, the solution inside the empty capsid has negative pressure similarly to the xylem of trees which would come

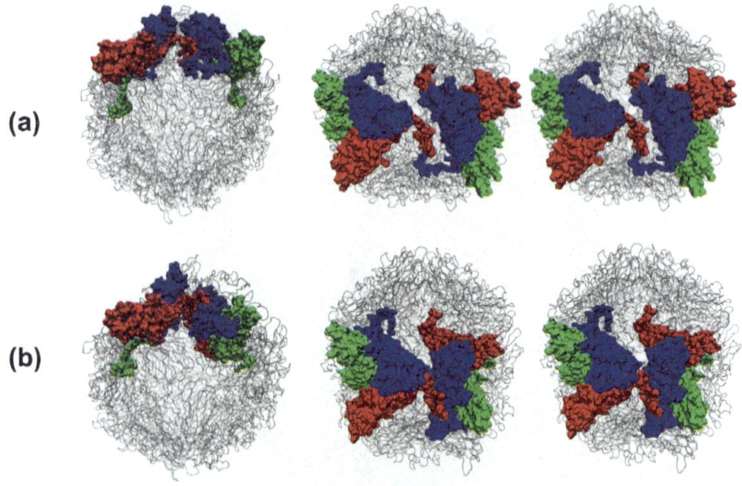

Fig. 4.50 Snapshot of collapse of satellite tobacco mosaic virus (STMV) capsid in the absence of RNA. **a** The beginning and **b** after 10 ns of the simulation. Trimers of the capsid heavily concerning the collapse are colored. Reprinted from Freddolino et al. (2006). Copyright 2006, with permission from Elsevier

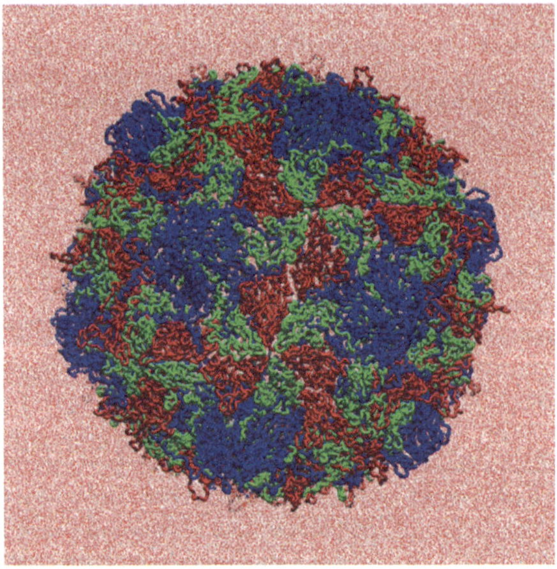

Fig. 4.51 The MD snapshot of the capsid of poliovirus in solution with coloring for VP1-VP4. Reprinted from Andoh et al. (2014), with the permission of AIP publishing

from the excessive charges in the capsid. This peculiar behavior seems to realize the environment to stably accommodate the RNA inside the empty capsid. The challenge of these MD calculations will be carried on for the time being until more sophisticated QM/MM or pure QM calculations become realistic.

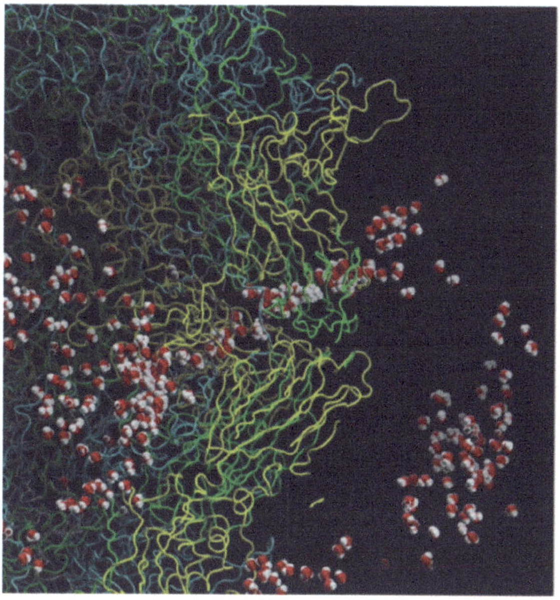

Fig. 4.52 The MD picture of exchanging water molecules between the inside and the outside of poliovirus under equilibrium. Reprinted from Andoh et al. (2014), with the permission of AIP publishing

References

Y. Andoh, N. Yoshii, A. Yamada, K. Fujimoto, H. Kojima, K. Mizutani, A. Nakagawa, A. Nomoto, S. Okazaki, J. Chem. Phys. **141**(165101), 1–10 (2014)

V. Aravindan, J. Gnanaraj, Y.-S. Lee, S. Madhavi, J. Mater. Chem. A **1**, 3518–3539 (2013)

B. Biswas, M. Sugimoto, S. Sakaki, Organometallics **19**, 3895–3908 (2000)

M. Brandbyge, J.-L. Mozos, P. Ordejón, J. Taylor, K. Stokbro, Phys. Rev. B **65**(165401), 1–17 (2002)

S. Datta, *Quantum Transport: Atom to Transistor* (Cambridge University Press, Cambridge, 2005)

R. Di Felice, A. Selloni, E. Molinari, J. Phys. Chem. B **107**, 1151–1156 (2003)

P.L. Freddolino, A.S. Arkhipov, S.B. Larson, A. McPherson, K. Schulten, Structure **14**, 437–449 (2006)

G. Frenking (ed.), *Topics in Organometallic Chemistry 12, Theoretical Aspects of Transition Metal Catalysis* (Springer, Berlin, 2005)

H. Fueno, M. Hayashi, K. Nin, A. Kubo, Y. Misaki, K. Tanaka, Curr. Appl. Phys. **6**, 939–942 (2006)

H. Fueno, Y. Kobayashi, K. Tanaka, Sci. China Chem. **55**, 796–801 (2012)

H. Fueno, S. Omae, and K. Tanaka, The 95th CSJ Annual Meeting, 1S4-07, Funabashi, Japan, Mar. 26 (2015)

S.I. Hauck, K.V. Lakshmi, J.F. Hartwig, Org. Lett. **1**, 2057–2060 (1999)

A. Ito, A. Taniguchi, T. Yamabe, K. Tanaka, Org. Lett. **1**, 741–743 (1999)

A. Ito, Y. Ono, K. Tanaka, Angew. Chem. Int. Ed. **39**, 1072–1075 (2000)

A. Ito, H. Ino, K. Tanaka, Polyhedron **28**, 2080–2086 (2009)

A. Ito, Y. Yamagishi, K. Fukui, S. Inoue, Y. Hirao, K. Furukawa, T. Kato, K. Tanaka, Chem. Commun., 6573–6575 (2008)

S. Iwamoto, Y. Inatomi, D. Ogi, S. Shibayama, Y. Murakami, M. Kato, K. Takahashi, K. Tanaka, N. Hojo, Y. Misaki, Beilstein J. Org. Chem. **11**, 1136–1147 (2015)

A. Kaji, T. Saeki, A. Ito, K. Tanaka, O. Hatozaki, The 52nd Battery Symposium in Japan, Tokyo, Nov. 27–29 (2011)

M. Koentopp, C. Chang, K. Burke, R. Car, J. Phys.: Condens. Matter, **20**(083202), 1–21 (2008)
K. Koike, S. Naito, S. Sato, Y. Tamaki, O. Ishitani, J. Photochem. Photobiol. A; Chem. **207**, 109–114 (2009)
Y. Kuramochi, M. Kamiya, H. Ishida, Inorg. Chem. **53**, 3326–3332 (2014)
R. Kuriki, K. Sekizawa, O. Ishitani, K. Maeda, Angew. Chem. Int. Ed. **54**, 2406–2409 (2015)
Y. G. Kuznetsov, S. Daijogo, J. Zhou, B. L. Semler, A. McPherson, J. Mol. Biol. **347**, 41–52 (2005)
J.-M. Lehn, R. Ziessel, J. Organometal. Chem. **382**, 157–173 (1990)
C.D. Lindstrom, M. Muntwiler, X.-Y. Zhu, J. Phys. Chem. B **109**, 21492–21495 (2005)
N. Mataga, Theoret. Chim. Acta (Berl.) **10**, 372–376 (1968)
N. Miyaura, K. Yamada, A. Suzuki, Tetrahedron Lett. **36**, 3437–3440 (1979)
I. Moritani, Y. Fujiwara, Tetrahedron Lett. **12**, 1119–1122 (1967)
M.K. Nazeeruddin, A. Kay, L. Rodicio, R. Humphry-Baker, E. Müller, P. Liska, N. Vlachopoulos, M. Grätzel, J. Am. Chem. Soc. **115**, 6382–6390 (1993)
H. Ohtani, Y. Hirao, A. Ito, K. Tanaka, O. Hatozaki, J. Therm. Anal. Calorim. **99**, 139–144 (2010)
J.P. Perdew, A. Zunger, Phys. Rev. B **23**, 5048–5079 (1981)
R.A. Periana, D.J. Taube, S. Gamble, S. Taube, T. Satoh, E. Fujii, Science **280**, 560–564 (1998)
A.E. Reed, L.A. Curtiss, F. Weinhold, Chem. Rev. **88**, 899–926 (1988)
J. R. Rumble ed., CRC Handbook of Chemistry and Physics, 99th Ed. 2018–2019 (CRC Press, Boca Raton, Fl 2018)
D. Sakamaki, A. Ito, K. Furukawa, T. Kato, K. Tanaka, Chem. Commun., 4524–4526 (2009)
S. Sakaki, Y. Ohnishi, H. Sato, Chem. Record **10**, 29–45 (2010)
J.M. Soler, E. Artacho, J.D. Gale, A. Garacía, J. Junquera, P. Ordejón, D. Sánchz-Portal, J. Phys.: Condens. Matter **14**, 2745–2779 (2002)
K. Stokbro, J. Taylor, M. Brandbyge, H. Guo, Lecture Notes Phys., **680**, 117–151 (2005) (G. Cuniberti, G. Fagas, and K. Richter, eds.) Springer, Berlin
M. Tachibana, K. Yoshizawa, A. Ogawa, H. Fujimoto, R. Hoffmann, J. Phys. Chem. B **106**, 12727–12736 (2002)
H. Takeda, K. Koike, H. Inoue, O. Ishitani, J. Am. Chem. Soc. **130**, 2023–2031 (2008)
K. Tasaki, A. Goldberg, M. Winter, Electrochim. Acta **56**, 10424–10435 (2011)
J. Taylor, H. Guo, J. Wang, Phys. Rev. B **63**(245407), 1–13 (2001)
M.M. Thackeray, C. Wolverton, E.D. Isaacs, Energy Environ. Sci. **5**, 7854–7863 (2012)
M. Winter, Z. Phys. Chem. **223**, 1395–1406 (2009)
L.D. Xing, J. Vatamanu, O. Borodin, G.D. Smith, D. Bedrov, J. Phys. Chem. C **116**, 23871–23881 (2012)
L. Xing, X. Zheng, M. Schroeder, J. Alvarado, A. von Wald Cresce, K. Xu, Q. Li, W. Li, Acc. Chem. Res., **51**, 282–289 (2018)
K. Xu, A. von Cresce, J. Mater. Chem., **21**, 9849–9864 (2011)
J. Yamada, T. Sugimoto (eds.), *TTF Chemistry: Fundamentals and Applications of Tetrathiafulvalene* (Kodansha Ltd, Tokyo and Springer, Berlin, 2004)
S. Yanase, T. Oi, J. Nucl. Sci. Technol. **39**, 1060–1064 (2002)
A. Yonezawa, T. Watanabe, M. Yokokawa, M. Sato, K. Hirao, in Proceedings of International Conference for High Performance Computing, Networking, Storage, and Analysis (Assoc. for Computing Machinery, Washington DC, USA, 2011), pp. 1–8
J. Zheng, D. Wang, W. Thiel, S. Shaik, J. Am. Chem. Soc. **128**, 13204–13215 (2006)

The manufacturer's authorised representative in the EU is Springer Nature Customer Service Centre GmbH, Europaplatz 3, 69115 Heidelberg, Germany. If you have any concerns regarding our products, please contact ProductSafety@springernature.com

Printed and bound by CPI Group (UK) Ltd, Croydon, CR0 4YY
30/09/2026
02225559-0002